T0213811

Springer Tracts in Civil Engineering

Series Editors

Sheng-Hong Chen, School of Water Resources and Hydropower Engineering, Wuhan University, Wuhan, China

Marco di Prisco, Politecnico di Milano, Milano, Italy

Ioannis Vayas, Institute of Steel Structures, National Technical University of Athens, Athens, Greece

Springer Tracts in Civil Engineering (STCE) publishes the latest developments in Civil Engineering - quickly, informally and in top quality. The series scope includes monographs, professional books, graduate textbooks and edited volumes, as well as outstanding PhD theses. Its goal is to cover all the main branches of civil engineering, both theoretical and applied, including:

- Construction and Structural Mechanics
- Building Materials
- Concrete, Steel and Timber Structures
- Geotechnical Engineering
- Earthquake Engineering
- Coastal Engineering; Ocean and Offshore Engineering
- Hydraulics, Hydrology and Water Resources Engineering
- Environmental Engineering and Sustainability
- Structural Health and Monitoring
- Surveying and Geographical Information Systems
- Heating, Ventilation and Air Conditioning (HVAC)
- Transportation and Traffic
- Risk Analysis
- Safety and Security

Indexed by Scopus

To submit a proposal or request further information, please contact:

Pierpaolo Riva at Pierpaolo.Riva@springer.com (Europe and Americas) Wayne Hu at wayne.hu@springer.com (China)

More information about this series at http://www.springer.com/series/15088

Kristian Dahl Hertz · Philip Halding

Sustainable Light Concrete Structures

 Springer

Kristian Dahl Hertz
Broenshoej, Denmark

Philip Halding
Department of Civil Engineering
Technical University of Denmark
Kongens Lyngby, Denmark

ISSN 2366-259X   ISSN 2366-2603   (electronic)
Springer Tracts in Civil Engineering
ISBN 978-3-030-80502-9   ISBN 978-3-030-80500-5   (eBook)
https://doi.org/10.1007/978-3-030-80500-5

This Springer imprint is published by the registered company Springer Nature Switzerland AG
The registered company address is: Gewerbestrasse 11, 6330 Cham, Switzerland

Preface

Light concrete structures offer solutions to many of the urgent problems builders and structural engineers are facing today. The small weight means less consumption of resources and possibilities of making longer spans. The open microstructure makes it fire-resistant and creates a sound damping effect. In addition, it is capable of regulating moisture content of enclosures. You may cast light concrete structures in textile moulds made with a minimum of materials in many spectacular shapes, since the pressure on the mould is small.

Super-light structures and sandwich structures use light concrete to fill out the space between stronger concrete parts, so that the light concrete can stabilize the strong and protect it from fire. Light concrete has a smaller E-modulus than the strong, and thereby the engineer can guide forces to follow an optimized path via the strong concrete parts. This means that the engineer decides where the forces should be in the structure. This is the opposite of the traditional approach, where the structure determines what the engineer must do, because the engineer should first find the path of the forces, before assessing the load-bearing capacity.

Super-light prefabricated SL-deck elements offer a large flexibility regarding shape and holes for pipes and implemented services as floor heating, sewers, water and electricity.

The combination of strong and light concrete makes it possible to establish a sound insulation with a material consumption of only three-fourth of what is elsewise required.

Prefabricated light floor systems allow beamless solutions with larger spans for offices, industry and car parks.

You may also utilize the super-light principle with combined application of light and strong concrete for creating tunnels and especially floating tunnels.

Another specific application is for shipbuilding, where the lightness of the structure improves manoeuvrability and reduces costs and resource demands compared to traditional concrete ships.

With pearl-chain technology, you can create curved shapes from simple flat mass-produced slab elements, which are easy and cheap to cast and transport to the erection site. This can reintroduce arches and vaults, which elsewise have become too difficult and costly to build.

Our new research in the lightest concrete qualities of densities down to about fifty kilogram per cubic metre has developed materials with insulating properties equal to mineral wool that can be cast out to fit anywhere.

Finally, modern builders and engineers are concerned about the impact on the climate.

They want to reduce the CO_2 emission from material production and building processes.

The reduced weight of super-light structures means less consumes of cement and aggregates, and in addition, we can now produce cement with less pollution and a smaller CO_2 impact. These numbers depend on who are producing the materials and how. However, the durability and long lifetime of the structures make them sustainable and in many cases more sustainable than, e.g., structures in wood.

It is not new to apply light concrete structures. Roman builders developed remarkably skills creating them. Roman concrete structures and technology represent a source of inspiration for modern builders and engineers. Their ideas and detailing have proved to be applicable in full scale and just wait for being reintroduced in combination with modern building technology. Super-light structures can be seen as examples of that. We therefore introduce the ancient light concrete structures in this book based on intensive studies on the subject.

The authors have written the present book in order to present the many technologies and detailed solutions available for light and super-light concrete structures. It is the intention that the book can serve as a design guide, a reference and a textbook for teaching the subject. In addition, the book is intended to serve as a source of inspiration for consulting engineers, architects and others interested in the subject and in creating a sustainable future.

Broenshoej, Denmark   Kristian Dahl Hertz
Kongens Lyngby, Denmark   Philip Halding

Acknowledgement

We would like to express our gratitude to the Realdania association for supporting the writing of the present book.

Contents

About the Authors

Kristian Dahl Hertz M.Sc., Ph.D. is a professor in design of building structures at the Department of Civil Engineering at the Technical University of Denmark (DTU).

After his Ph.D. on Properties of Fire Exposed Concrete from 1980, he was a part-time associate professor at DTU and a part-time consulting engineer for M. Folmer Andersen Ltd. Here, he designed building structures for the Ministry of Foreign Affairs of Saudi Arabia, the Royal Theatre in Copenhagen, the National Bank of Iraq, the Parliament of Greenland, factories and domestic buildings.

Kristian began teaching in Structural Fire Safety Design and new courses in structural building design at DTU. He made research on these subjects in cooperation with consulting engineers and producers of light aggregate concrete elements. For 16 years, he was a regular visiting professor at London City University. During the 1990s, he also served as a head of department and merged 14 small DTU departments into a Department of Civil Engineering. At the same time, he was active in code writing and implemented some of his design methods in Danish and European codes.

In 2009, Kristian invented super-light structures and pearl-chain technology and established a course and a special research area on sustainable light concrete structures at DTU.

In 2010, three young business persons asked him to make a spinout company Abeo Ltd based on the ideas, and the same year, the company won a world championship in clean technology in San Francisco (Clean Tech Open).

A cooperation between DTU, the company, test institutions, producers, architects, consulting engineers and contractors developed the new technology and many practical solutions and details.

They applied the findings in a large number of buildings and structures. Factories now produce super-light concrete elements in the USA and more European countries.

In 2019, Kristian collected most of the findings about fire safety of concrete structures in the book "Design of Fire-resistant Concrete Structures." In 2020 and 2021, he and Philip Halding collected the findings on light concrete structures, super-light structures, and pearl-chain technology in the present book.

Philip Halding M.Sc., Ph.D. is an assistant professor at the Technical University of Denmark (DTU) in the Department of Civil Engineering. He is specialized in sustainable super-light structures and full-scale testing and monitoring of concrete bridges.

In his Ph.D., he worked with the pearl-chain method used for bridges, and he was part of the development from a simple theoretical concept until the erection of the first-ever pearl-chain bridge. He cooperated with several industrial partners and another Ph.D. with focus on the types of concrete used.

After his Ph.D., Philip has worked with advanced monitoring methods of existing concrete bridges during full-scale load tests. The purpose of such tests was to upgrade existing bridges for heavier traffic by wish from the Danish Road Directorate. The monitoring methods included non-intrusive wide-angle digital image correlation (DIC), laser scanning and interferometric radar, as well as a number of other types of equipment.

Philip is the primary teacher of the two DTU master courses: "super-light structures" and "structural analysis of buildings." The course in super-light structures involves about 80 students in group work with the different concepts and ideas presented in this book.

Furthermore, Philip is supervising numerous bachelor and master student project every semester with focus on super-light technology. Some of the projects are in collaboration with the Abeo company, so that the students work in close collaboration with the industry. Most of the projects concern actual real-life challenges, where a builder wishes to create a more aesthetical or sustainable solution, or where the project problem is part of the current research within super-light structures.

In addition, Philip has graduated the "Pasteur program," which is a project management course from Harvard Business School, and he works actively with state of the art in teaching methods at university level.

Philip has over the years worked together with Kristian, first with Kristian as a Ph.D. supervisor and later as colleagues. A large part of the present book is based on this fruitful research collaboration in the past decade.

Notations

A	Area
A	Area of openings of a compartment
A_1-A_5	Life cycle phases
A_c	Area of concrete
A_{lc}	Area of light concrete
A_s	Area of steel
A_{tot}	Total enclosing surface area of a compartment
A_{ts}	Area of crossing steel
B_1-B_5	Life cycle phases
C_1-C_4	Life cycle phases
D	Diameter
D	Life cycle phase for reuse
D_c	Diameter of strong concrete
D_{lc}	Diameter of light concrete
D_m	Outer diameter of concrete with entasis
D_s	Diameter of steel
E	E-modulus
E_0	Initial E-modulus
E_{0a}	Initial E-modulus of a concrete number a
E_{0b}	Initial E-modulus of a concrete number b
E_a	E-modulus of a concrete number a
E_b	E-modulus of a concrete number b
E_c	E-modulus of concrete
E_{c0}	Initial E-modulus of concrete
E_{c020}	Initial E-modulus of concrete at 20 °C
E_{lc}	E-modulus of light concrete
E_{lc0}	Initial E-modulus of light concrete
E_s	E-modulus of steel
E_{s0}	Initial E-modulus of steel
E_{s020}	Initial E-modulus of steel at 20 °C
EA	Axial stiffness
EA_0	Initial axial stiffness

EI	Flexural stiffness
EI_0	Initial flexural stiffness
EI_{c0}	Initial flexural stiffness of concrete
EI_{lc0}	Initial flexural stiffness of light concrete
EI_{s0}	Initial flexural stiffness of steel
F	Force
F_0	Strength of centrally loaded elastic column with entasis
F_E	Euler force
F_R	Rankine force
F_{R0}	Rankine force of column with entasis
F_{cE}	Euler force of concrete in a column
F_{cR}	Rankine force of concrete in a column
F_{cc}	Strength of concrete in a cross section
F_{cu}	Ultimate force of concrete in a cross section
F_{ibot}	Ring force at bottom of level i of a cupola
F_{itop}	Ring force at top of level i of a cupola
F_{lc}	Strength of light concrete in a cross section
F_{lcE}	Euler force of light concrete in a column
F_{lcc}	Strength of light concrete in a cross section
F_{lcu}	Ultimate force of light concrete in a cross section
F_{res}	Resulting ring force at a level of a cupola
F_s	Force in steel
F_{s1}	Force in steel layer 1
F_{s2}	Force in steel layer 2
F_{sE}	Euler force of steel in a column
F_{su}	Ultimate force of steel in a cross section
F_u	Ultimate force (ultimate resistance of a cross section)
H	Height of an arch
H_i	Height at level i of a cupola
I	Moment of inertia
I_c	Moment of inertia of concrete
I_{lc}	Moment of inertia of light concrete
I_s	Moment of inertia of steel
K	Parameter
L	Length
M	Moment
M_{End}	Moment at end of a beam
M_c	Moment in concrete
M_i	Moment on an element at level i of a cupola
M_s	Moment in steel
N	Normal force
O	Opening factor
P	Single load
P_i	Load at level i
R	Radius

T	Temperature
T_1-T_{64}	Temperature parameters
V_i	Horizontal reaction at level i
a	Parameter
b	Parameter
b	Parameter for column calculation
b_1	Throat height of concrete hinge
b_2	Height of concrete hinge
c_c	Width of concrete cross section
c_{lc}	Width of light concrete cross section
c_p	Heat capacity
c_s	Cover to steel axis
d	Cover thickness
d_F	Depth of force F
d_i	Horizontal distance to load from inclined cupola element
d_{s1}	Depth of steel layer 1
d_{s2}	Depth of steel layer 2
e	Eccentricity
f_{cc}	Compressive strength of concrete
f_{cc20}	Compressive strength of concrete at 20 °C
f_{ct}	Tensile strength of concrete
f_{cu}	Ultimate strength of concrete
f_{lcu}	Ultimate strength of light concrete
f_s	Yield strength of steel
f_{s20}	Yield strength of steel at 20 °C
f_{su}	Ultimate strength of steel
f_u	Ultimate strength
h	Height of cross section
h	Height of opening
h	Height to a point on a compression arch
h_c	Height of concrete cross section
h_{cupola}	Height of center of cupola
h_{lc}	Height of light concrete cross section
k	Factor for bond shear strength as part of compressive strength
k	Remaining part of strength after cooling to high temperatures
m	Parameter
p	Load per unit length
q	Fire load per unit enclosing surface area of compartment
r_i	Horizontal radius at level i of a cupola
t	Time in minutes
t_c	Thickness of concrete flange
v	Angle parameter
w	Width
w_{itop}	Width at top of cupola element at level i
x	Length parameter

y	Length parameter
y_{ca}	Height to catenary arch
y_{cc}	Length to concrete edge in compression
y_{ci}	Height to circle arch
y_{ct}	Length to concrete edge in tension
y_{pa}	Height to parabola arch
y_s	Length to steel axis
α	Thermal elongation coefficient
α	Angle
δ	Deflection
δ_u	Ultimate deflection
ε	Strain
ε_F	Strain at a force F
ε_{c20}	Strain of concrete at 20 °C
ε_{cu}	Ultimate strain of concrete
ε_s	Strain of steel
ε_{s20}	Strain of steel at 20 °C
ε_{su}	Ultimate strain of steel
ε_{su20}	Ultimate strain of steel at 20 °C
ε_{sy}	Yield strain of steel
ε_u	Ultimate strain
η	Stress distribution factor
κ	Curvature
λ	Conductivity
ν	Poisson's ratio
ξ	Reduction of strength
ξ_c	Reduction of concrete strength
ξ_{ccCOLD}	Reduction of compressive strength of concrete in cold condition
ξ_{ccHOT}	Reduction of compressive strength of concrete in hot condition
ξ_{cM}	Reduction of concrete strength at midpoint
ξ_s	Reduction of steel strength
ρ_c	Density of concrete
σ	Stress
σ_I	Biaxial stress in direction I
σ_{II}	Biaxial stress in direction II
σ_b	Bearing stress
σ_{c1}	Concrete stress at side 1
σ_{c1m}	Concrete stress at side 1 at middle of column
σ_{c1top}	Concrete stress at side 1 at top of column
σ_{c2}	Concrete stress at side 2
σ_{c20}	Concrete stress at 20 °C
σ_{c2m}	Concrete stress at side 2 at middle of column
σ_{c2top}	Concrete stress at side 2 at top of column
σ_{cau}	Biaxial concrete strength with hindered lateral expansion
σ_{itop}	Horizontal stress at top of level i of a cupola

σ_{s20} Steel stress at 20 °C
σ_{ts} Steel stress 31.5 MPa at ultimate tensile strain of concrete
τ Shear stress

Chapter 1
History

Abstract The historical development of concrete and light concrete is explained and dates all the way back to the Phoenicians at approximately 1200 BC. Super-light concrete technology is also inspired by techniques of vaults and cupolas developed in the Roman Empire.

1.1 Ancient Building Materials

1.1.1 Gypsum and Limestone Mortar

Now and then, you hear a theory that the pyramids of Giza in Egypt (2580 BC) were made of concrete. That is not true. The largest of them are made of blocks of limestone placed in a mortar that according to the author's investigations primarily is made of gypsum with small grains of burned clay. An orange-red colour of the grains caused by the content of iron oxide shows that the material has been burned at 400 °C [1, 2], where less than 300 °C is needed for burning a gypsum mortar [3].

This means that the Egyptians, for whom wood was a limited resource, could produce gypsum mortar with less energy than needed for burning limestone (900 °C) and for making a cement. Furthermore, builders were familiar with gypsum at that time, since you can trace it as a building material to 4.000 BC in the Sumerian city Uruk [4] and perhaps even back to 9.000 BC in Catalhöyük in Turkey [5].

However, the pyramids can tell us something else of importance for the later concrete production. If you measure their orientation, you will find that all 80 pyramids in the Northern Egypt are rotated about four degrees with respect to North (Fig. 1.1).

North is in the middle between the direction where the sun rises and where it sets, so the Egyptians had no problems in determining that. (Explanation follows).

When the desert developed in the northern Africa and Middle East, people had to concentrate at the rivers. The first strong societies therefore appeared here, and the rulers needed a way to demonstrate their power.

King Djoser's architect Imhoteph placed the first step pyramid in Saccara at the Nile about 2660 BC (Fig. 1.2). He probably intended to make a copy of mountains

© The Author(s), under exclusive license to Springer Nature Switzerland AG 2022
K. D. Hertz and P. Halding., *Sustainable Light Concrete Structures*, Springer Tracts in Civil Engineering, https://doi.org/10.1007/978-3-030-80500-5_1

Fig. 1.1 Pyramids in Giza. *Photo* KD Hertz

Fig. 1.2 King Djoser's step pyramid. *Photo* KD Hertz

with the same shape, which serve as landmarks.

(You may for example find one in the desert East of Oman shown in the film "Lawrence of Arabia", taken on location in 1962).

Wind from North and West have eroded soft layers of these natural mountains, so that they became step pyramids similar to the artificial one he built.

We can also measure this first pyramid to be rotated four degrees with respect to North.

The natural preconditions for developing new building materials were better in other areas of the ancient world.

The Phoenicians lived for example in Lebanon and northern Israel and had rich amounts of fuel from the forests. They burned limestone and clay for mortar, tiles, and bricks and they had a large production of glass. They exported glass pearls, vases, and raw glass bars to the entire Mediterranean.

1.1.2 Ancient Concrete

The Phoenicians made the oldest concrete that the authors know about approximately 1200 BC. As mentioned above, they had a large glass production (Fig. 1.3). They pulverized glass and mixed it with burned limestone and water creating calcium-silicate-hydrate. A material, which we know as hydrated cement. They also made cement from pulverized tiles and burned limestone.

The Bible tells us that the Phoenician king gave King David some masons, when he founded Jerusalem. That was not just a helping hand. It was technology transfer introducing concrete as a new building material. Today, you can see concrete claddings of water cisterns made from this material in Jerusalem [5].

Later, the Phoenicians learned to use fly ash from the volcano at Santorini replacing the more expensive pulverized glass and tile in their concrete. At this island, you can see Phoenician fly ash quarries next to Roman quarries showing us how the Romans may have learned about concrete from the Phoenicians.

The Romans identified fly ash applicable for concrete production from volcanoes at many locations in their empire. Vitruvius talks about fire in the earth [6].

Fig. 1.3 Phoenician glass. *Photo* KD Hertz

This has something to do with the orientation of the pyramids.

The reason why we have these volcanoes on a line in the direction East–West through the Mediterranean is that the whole continent of Africa has turned four degrees since they were built. With Gibraltar as centre of rotation, the African tectonic plate dives under the European creating a basis for volcanic activity. It is amazing that we can read the movement of continents on man-made structures.

In Italy, the Romans found plenty of fly ash especially from the European super-volcano Campi Flegrei (Fig. 1.4) at Pozzuoli near Naples. This city has given the name Pozzolana to the siliceous materials reacting with calcium oxide (burned limestone) and water to calcium silicate hydrate (hydrated cement). Big eruptions of this super volcano at 35.000 BC and 10.000 BC gave volcanic ash especially to the East as far as in southern Russia.

This and materials from other volcanoes became a basis for the Roman concrete production.

In southern Germany, they found tuff that was applicable. In areas with none or sparse occurrence of natural Pozzolana, they applied the old recipes based on pulverized tiles.

They could also mix this with natural fly ash. Furthermore, they mixed cement with aggregates of stones, pieces of rock, or crushed parts of old concrete or tile structures to make the new concrete also known as opus caementicium.

Fig. 1.4 Campi Flegrei. From the crater of the European super volcano. *Photo* KD Hertz

1.2 Ancient Structures

1.2.1 Walls

For walls and columns, the Romans cast concrete between surface shells of bricks or stones, which they masoned simultaneously with casting the concrete.

Besides from participating in the load-bearing function of this composite structure, the shells served as permanent moulds and as endurable surfaces protecting the concrete against mechanical impact, weather, and heat from fire.

Roman builders applied mainly three different kinds of permanent moulds.

Opus Incertum consists of irregular stones placed as the plane surfaces keeping the concrete core in place. The Roman wall next to Tower of London is an example of that.

For nice surfaces, Roman builders applied Opus Latericium (Fig. 1.5) that is masonry of flat bricks of burned clay.

They often gave the flat bricks a pentagonal shape with a rectangular part at the surface and a triangular part pointing into the central concrete material to obtain a better grip. You typically find Opus Latericium in facades and on ends of internal walls, where they become visible.

They used Opus Reticulatum as shown in Fig. 1.6 mainly for internal walls.

It was fast to build, because the builder could easily place the long pyramidal ceramic tiles with quadratic bottom in the concrete and simultaneously adjust skewness and unevenness.

Fig. 1.5 Opus Latericium. *Photo* KD Hertz

Fig. 1.6 Opus Reticulatum. *Photo* KD Hertz

When Phoenicians and Romans applied fly ash from the volcanoes in their concrete, it is only natural that they also applied volcanic stones including pumice as aggregate.

This means that they applied what we today call light aggregate concrete and they often used several concrete qualities in the same structure to obtain a suitable distribution of weight and to guide the forces in the structure to where they wanted them to be and to avoid tension as far as possible [7, 8].

Application of light aggregate concrete reduces the dead load and thereby the pressure on the permanent moulds. As mentioned, they could also apply pieces of tile as aggregate, and sometimes they used hollow pots or vases as aggregate.

By means of this, they were able to include larger air filled volumes in order to obtain a lower weight and density of the concrete structure than possible by application of natural pumice as light aggregate.

The pots of burned clay also had a certain strength, which means that the master builder obtained a weight reduction without suffering the same strength reduction as would follow if he applied a natural pumice aggregate to get the same low density.

1.2.2 Arches and Vaults

Builders have applied arches since pre historic time. Arches are capable of withstanding uniformly distributed load in a perfect way (without bending moments). The main benefit is that you may obtain a certain span only by application of compression forces, which you can guide from load to support by means of unreinforced structures of stone and concrete.

In the Egyptian museum in Cairo, you can see model toy houses of burned tile from 2500 BC, where the buildings are made of vaults, which give lateral support to each other (Fig. 1.7).

You could imagine that the ancient builders made their multi vault structures like present day builders do it in Sudan. Figure 1.8 shows an example of the technique.

First, they make the walls of the houses and their gables. Then, they place mud bricks, fibre reinforced with straw, as inclined arches leaning against the gable and

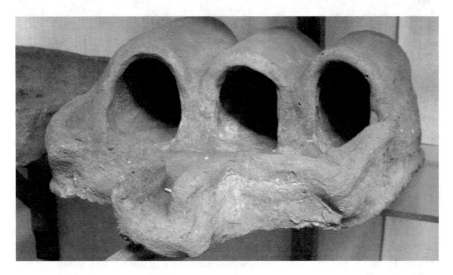

Fig. 1.7 Egyptian toy house 2500 BC. *Photo* KD Hertz

Fig. 1.8 Vaults in Sudan made without scaffolding. *Photo* KD Hertz

Fig. 1.9 Granaries in Ramesseum in Luxor 1400 BC. *Photo* KD Hertz

therefore without any scaffolding! Finally, they finish the vault with an outer mortar, or they may apply a second layer of brick arches leaning against the opposite gable.

The builders applied the same basic vault building principles in the granaries of the Ramesseum in Luxor from about 1400 BC (Fig. 1.9). Here the approximately 100 m long vaults also support each other laterally. Each vault span about 8 m and consists of four layers of mud bricks, which are fibre reinforced with straw.

Roman builders developed arch building technology and applied it for many purposes.

In a bridge like Ponte Fabricio (Fig. 1.10) the master builder made two free spans of 24.5 m to reach the Tiber island—Isola san Bartolomeo—in Rome.

The bridge is from year 62 BC, and history tells us that the master builder had to wait 20 years for his payment, because the authorities would like to be sure that it could stand the impact of the river.

Roman builders applied concrete vaults covering paths and staircases in most public buildings.

They also applied vaults of considerable dimensions for side bay of basilicas.

For example, a full symphony orchestra can find place in one of the six side bays of the Basilica of Maxentius and Constantin in Rome.

One important application of the ancient vault is as fire protection.

A vault carries its load by leading compressive forces through fire resistant materials as concrete and brick. A possible shrinkage or deformation of these materials may cause a minor settlement but not a major collapse.

Roman builders therefore applied them for ceilings over shops at the bottom of their buildings separating the combustible content of the shops from the upper stories, where other people lived, and where they might apply lighter and more flammable wooden floors.

Fig. 1.10 Ponte Fabricio. *Photo* KD Hertz

Isolated staircases of stone, brick, or concrete lead to the upper stories.

The builders also usually installed a wooden floor under the vault of the shop, so that the owner's family could sleep there and serve as a living fire alarm.

Figure 1.11 shows an elegant Roman concrete vault covering a circular path. Here, the master builder has created a vault with an extremely small thickness of only few centimetre, and constructed a horizontal ring beam where the lateral forces from the

Fig. 1.11 Ultra-thin concrete vault Villa Adriana. *Photo* KD Hertz

Fig. 1.12 Horizontal vault Trajan's Market. *Photo* KD Hertz

vault is counteracted by compression in the ring so that the columns should only carry vertical reactions.

Figure 1.12 shows another elegant application of a vault. Here the master builder (Apollodorus from Damascus) shaped the entire eight storey high building of Trajan's Market in Rome as a big vault resisting the lateral pressure from the mountain that was carved out in order to make space for the market.

1.2.3 Cupolas

Application of light aggregate concrete based on pumice reduces the dead load. The best Roman master builders took advantage of the light aggregate concrete in their shell structures (such as vaults and cupolas), where the reduction of the dead load opened possibilities for creating large span widths. As mentioned above, they could also mix pots of burned clay in the concrete in order to reduce the density by creating empty volumes.

Figure 1.13 shows a cross-section of a cupola from Hadrian's villa outside Rome.

As for many other cupolas, the master builder of this one constructed a horizontal round hole at the top, where the forces become small.

Sometimes, as for example in the Trajan's Market in Rome, a horizontal compression ring of bricks provides the round hole with a well-defined edge, and it serves to counteract a possible residue of radial compression forces.

The relatively thick walls and pillars beneath the cupola lead the horizontal reactions of the cupola further into the ground.

Fig. 1.13 Light concrete shell Villa Adriana. *Photo* KD Hertz

As also seen in Fig. 1.13, a series of arches support the cupola and lead the forces down in pillars between openings and niches in the wall below.

You may often find such relieving arches in walls beneath cupolas, even when you do not find openings or niches. This is for example seen several places in the walls supporting the cupola of Pantheon in Rome.

We know that the Romans could apply several layers of light concrete of different density in the same structure. They did for example apply no less than eight horizontal layers of light concrete with different densities in the construction of the Pantheon.

By means of that, they ensured that the compressive forces were kept within the cross-sections for the large span of 43 m [9].

Apollodorus of Damascus, who rebuilt the temple and constructed the present cupola, designed this light concrete shell with cassettes (deep panels with less material thickness, also known as coffers) (Fig. 1.14). By doing so, he further reduced the weight compared to its load bearing capacity and simultaneously, he guided the forces into the radial and circular ribs.

You find other delicate, load-bearing shell structures in the many half cupolas of the Roman baths. Today, we can only guess how they were able to design them. Modern engineers have a severe challenge if they try to construct similar buildings and estimate the stress distributions without the aid of computer software. (See 8.2.4).

1.2.4 Roman Reinforced Concrete Structures

The Roman builders applied concrete and light concrete in structures, where forces act in compression. Arches and vaults lead vertical loads to the sides where thick

Fig. 1.14 Pantheon Cupola. *Photo* KD Hertz

walls or massive sections withstood the lateral forces from the arches. The walls were hollowed out with niches or supported by cross walls.

Nevertheless, one particular structure called for application of flat concrete slabs subjected to bending: "Hypocaust"—a type of floor heating system.

They lead hot air under the floors in bath buildings, and from there the system could guide it up via ceramic pipes embedded in the walls.

For such floors, vaults were not so applicable because the height could be limited and because a vault would capture the hot air and hinder sideward's distribution of it.

Flat concrete slabs were therefore preferable.

This lead to a structural problem, since such flat slabs act in bending for which plain concrete cannot resist much tension stresses.

The Romans therefore in general had to support the flat concrete slabs with piles of brick at a small distance from one another as shown in Fig. 1.15.

However, the authors know about at least one structure showing another solution to this problem.

In a slab covering a floor-heating system in the so-called "Repräsentationshaus" at Magdalensberg near Klagenfurt in Austria, the Roman builder introduced an iron reinforcement to take tension forces in the bottom side [10].

Here, the builder placed a series of 1800 mm long flat iron bars with cross-section 5 times 30 mm and a 200 mm spacing to resist the tension forces in the bottom of a 160 mm thick concrete slab covering a 1070 mm wide hot air channel [5], (Fig. 1.16).

The Romans called the Klagenfurt area Noricum. It was famous for steel production.

Fig. 1.15 Roman hypocaust system. *Photo* KD Hertz

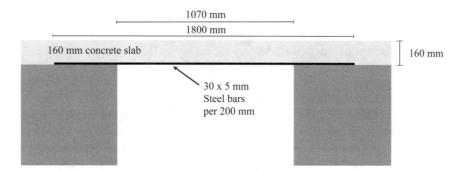

Fig. 1.16 Norican iron reinforcement

Each year merchants from the whole empire came here to buy tools and weapons of the best quality. It is therefore no wonder that we find the first steel reinforcement in a concrete structure here.

The Roman builders also applied fibre reinforcement to take tensile stresses or to distribute cracks from tensile forces. Fibre reinforcement could be many different qualities, but the builders often applied hairs from horses or donkeys.

Making aqueducts, builders had the task to keep the structure watertight, also when the terrain was setting or deforming. They could therefore apply up to eight different layers of mortar (opus signinum) in the channel.

In such hydraulic structures, builders could make the mortar near the water tight with olive oil. In order to avoid cracks in this dense layer, they could place one or more layers of mortar with fibre reinforcement that distributed cracks from the deeper layers.

Fig. 1.17 Aqueduct and pressure pipe at Aspendos in Turkey. *Photo* KD Hertz

The inmost mortars often contained red pulverized burned clay. The next layers often contained pulverized marble. Then followed layers with charcoal. This made a typical sequence of red, white, and grey mortar layers [11].

Sometimes the builders placed the water channel on a layer of small arches, which served two purposes. To create the inclination wanted and to distribute settings from the larger arches with longer spans below.

In some places, the builders applied pressure pipes to avoid building a high aqueduct for example across a wide valley as that in Aspendos. Here, it was a problem, that the dynamic inertia of the water flow might give a blow destroying the pipe if something interrupted the flow. Today, we know this effect as a "hydraulic ram".

At Aspendos, the master builder solved this problem by making the aqueduct in full height at two points as seen in the Fig. 1.17 and placing an open ventilated basin at the top of each.

In other locations, they solved the problem by providing pressure pipes with safety valves. They made these as two or three holes in the strong concrete pipe filled with a lighter concrete of less strength.

References

1. Hertz KD (1980) Betonkonstruktioners Brandtekniske Egenskaber. (Properties of Concrete Structures Exposed to Fire). (In Danish). Department of Building Design, Technical University of Denmark. 209p
2. Hertz KD (2019) Design of fire-resistant concrete structures. ICE-Publishing, Thomas Telford Ltd. ISBN: 9780727764447. 248p. London 2019

3. Eckel EC (1905) Cements, Limes and plasters. Their materials, manufacture, and properties. Wiley, New York. 698p
4. Haegermann G (1964) Vom Cæmentum zum Spannbeton, Part 1. (In German). Bauverlag GmbH, Wiesbaden-Berlin. 491p
5. Stark J, Wicht B (1998) Geschicte der Baustoffe. (In German). Bauverlag, Wiesbaden und Berlin 205p
6. Vitruvius (23 BC) The Ten Books on Architecture. By Morgan MH (1960) (In English) Dover New York, 331p, and by Isager J (2016) Om Arkitektur (In Danish) South Danish University Publishing, Odense, Danmark. 505p
7. Hertz KD (2010) A new patented building technology based on ancient Roman knowledge. In: Proceedings of symposium "handling exceptions in structural engineering" La Sapienza, 6 p Rome, Italy, 8–9 July 2010
8. Hertz KD, Schmidt JW, Goltermann P (2014) Super-light and pearl-chain technology for support of ancient structures. In: Proceedings of the 2'nd international conference on protection of historical constructions PROHITECH'14, pp 641–646, May 2014, Antalya, Turkey
9. de Fine Licht KR (1968) The Rotunda in Rome. Jutland Archaeological Society Publications, A Study on Hadrian's Pantheon, 344p
10. Lamprecht HO (1985) Opus Caementitium Bautechnik der Römer. (In German) Beton-Verlag Düsseldorf
11. Frontius SJ (98 AD) De Aquae Ductu Urbis Romae. (About the Water Supply of the City of Rome) Translated by Hansen J (In Danish) 1982 Museum Tusculanum 215p

Chapter 2
Materials

Abstract Super-light structures utilizes concrete types of different strengths and densities. Knowing the properties of the different concrete types can enable engineers to build lighter and with fire safety. The chapter also provides information about the use of reinforcement steel and pre-stressing.

In this chapter, we present data for a number of materials relevant for the structures in the book. We do that in order to give the reader an idea of which materials, we have in mind for the structures presented.

Designers can use the data for preliminary calculations until they choose the actual materials for their structures with respect to local conditions such as availability of natural minerals, price levels, production methods, and local codes, legislation, and traditions.

We recommend you to read about the history and origin of building materials in Chap. 1 first, because this may give you a deeper understanding of their nature and detailed composition.

Figures 2.1 and 3.3 show an example of an advanced application of reinforced concrete for an optimal structure by Piere Luigi Nervi [1, 2].

2.1 Heavy Concrete

2.1.1 Ordinary Concrete

The concept of concrete comprises a variety of materials with different aggregates, cement qualities, mixtures, strengths, densities and other physical properties.

Aggregates depend usually on available local resources as for example granite, basalt, limestone, and quartz. Producers can also apply other natural or artificial aggregates like pumice and ceramics.

Cement can vary for example depending on the local availability of lime. Producers and contractors can substitute it partly or fully by artificial or natural

© The Author(s), under exclusive license to Springer Nature Switzerland AG 2022
K. D. Hertz and P. Halding., *Sustainable Light Concrete Structures*, Springer Tracts
in Civil Engineering, https://doi.org/10.1007/978-3-030-80500-5_2

Fig. 2.1 Palazzetto dello sport by PL Nervi 1958. *Photo* KD Hertz

pozzolana like fly ash from volcanoes and chimneys or crushed ceramics, and they can apply a great variety of chemical additives influencing the physical properties and workability of the concrete.

For the purpose of this book, we need to identify some groups of concrete materials and apply them for different purposes. A heavy concrete has usually a density of at 2300 kg/m^3 or more. In this book, we call a concrete heavy, if the density is at least 2300 kg/m^2. We call all qualities of less density light. This definition accord with the distinction applied in most other codes and textbooks.

Some concrete codes confine themselves to certain intervals of strength. The Eurocode [3] for example apply an upper limit of strength of 60 MPa.

In practise, engineers often apply stronger materials, but treat them according to the codes by using a calculational strength within the limit instead of the real measured physical strength.

In this book, we call a heavy concrete, which is defined to be within 60 MPa, for ordinary concrete or just concrete in contradiction to a high-strength concrete (See Sect. 2.1.2).

In practise, you should find and document the material properties by tests. However, as a guideline for preliminary calculations and student projects, we give some typical values in Table 2.1. They are all based on 5% quantiles (95% of a series

Table 2.1 Concrete data

Compressive strength f_{cc}	10	20	30	40	50	60	MPa
Tensile strength f_{ct}	1.0	1.5	2.0	2.5	2.9	3.1	MPa
Initial E-modulus E_{c0}	22	31	36	38	40	42	GPa

of tests has higher strength or stiffness) from own not reported test results, supported by CEN [3], DS [4].

For many applications of heavy concrete in super-light elements dealt with in this book, we apply a concrete with $f_{cc} = 55$ MPa, $f_{ct} = 3.0$ MPa, $E_{c0} = 41$ GPa, thermal elongation coefficient $\alpha = 1.1 \cdot 10^{-5}$ /°C, heat capacity $c_p = 1.00$ kJ/m³, and conductivity $\lambda = 0.90$ W/m°C.

Most textbooks and codes on concrete structures apply idealized stress–strain curves for concrete. They often give formulas for the curves, which look quite different, because they are semi-empirical and apply additional old-fashioned assumptions such as a ratio 1000 between the initial E-modulus and the strength.

However, you can often identify the curves as derived from the same basic assumption made by Ritter. According to him, you can find the E-modulus of a concrete at a certain compressive stress as the initial E-modulus reduced by the ratio between the stress and the strength [5].

This means, that the E-modulus E_c at a certain compressive stress σ on a concrete with a compressive strength f_{cc} at an ultimate strain ε_{cu} (which is usually 0.35%) can be expressed as

$$E_c = E_{c0}(1 - \sigma/f_{cc}) \text{ or } d\sigma/d\varepsilon = E_{c0}(1 - \sigma/f_{cc})$$

Solving this differential equation, you find the following expression for the working curve

$$\sigma(\varepsilon) = f_{cc}(1 - e^{-(E_{c0}/f_{cc})\varepsilon})$$

Figure 2.2 shows the curve.

If the concrete is heated up for example by a fire exposure, aggregates expand and cement paste releases physically and chemically bound water and shrinks. The concrete will therefore become more porous, crack, and lose strength, stiffness and conductivity. These reductions will therefore mainly depend on the aggregate applied [6].

Fig. 2.2 Idealized stress–strain curve for concrete

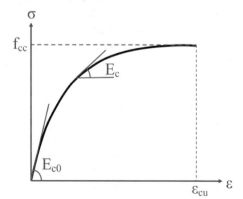

When the concrete cools down at the end of a fire, the cement paste will absorb water again, and new calcium hydrate crystals will widen up the cracks.

As a result, strength and stiffness will reduce further in the cooling phase. We therefore distinguish between the reduction $\xi_{ccHOT}(T)$ of the concrete compressive strength at a certain temperature T in a HOT condition during a fire and the final reduction $\xi_{ccCOLD}(T)$ in a COLD condition after the concrete has cooled down from a maximum temperature T.

See detailed explanations and methods for fire-safety design of concrete structures in the textbook "Design of Fire-resistant Concrete Structures" [7].

Strength reduction as a function of temperature (in °C) will be an S-shaped curve, and for all concrete and steel materials. You can calculate it by the same general formula

$$\xi(T) = k + \frac{1-k}{1 + \left(\frac{T}{T_1}\right) + \left(\frac{T}{T_2}\right)^2 + \left(\frac{T}{T_8}\right)^8 + \left(\frac{T}{T_{64}}\right)^{64}}$$

k is the part of a steel strength regained after cooling, and it is always 0 for concrete.

T_1. T_2, T_8, and T_{64} are constants in degree Celcius for the actual material. You find the values in Table 2.2. The E-modulus of a concrete will be reduced by the square of the strength reduction. This means that

$$f_{cc}(T) = \xi_c(T)f_{cc20} \quad , \quad E_{c0}(T) = \xi_c(T)^2 E_{c020} \quad \text{and} \quad \varepsilon_{cu}(T) = 0.35\%/\xi_c(T)$$

Any point $(\varepsilon_{c20}, \sigma_{c20})$ at the stress–strain curve is transferred to $(\varepsilon_{c20}/\xi_c(T), \xi_c(T) \cdot \sigma_{c20})$.

The temperature in a HOT condition and the maximum temperature during the fire, which you apply in a COLD condition, vary through a concrete cross-section. You can consider this variation by using a reduced cross-section, where you apply the actual strength reduction ξ_{cM} in the midpoint. You reduce the width by subtraction of damaged zones at the fire-exposed surfaces. The reduction factor η for the width represents the average reduction of strength related to that in the midpoint [7]. We call the value η the stress-distribution factor. It is between 0 and 1.0.

You can find free programs for the strength reduction ("Damage") and for the temperature distribution and average strength reduction ("ConFire") at the homepage of the Department of Civil Engineering [8], where you also find User's guides for the programs.

The program ConFire calculates temperature and strength reduction of a rein-forcing bar or a concrete in a specific point of a cross-section at a specific time t in minutes. This is based on the time of max temperature during a fire (the HOT condition) and the max temperatures during a full fire course (the COLD condition).

The fire courses can be a standard fire according to [9] or a fully developed fire. The latter is defined by a fire load q in MJ per m^2 enclosing surface A_{tot}, and an opening factor O in m^{½} depending on the areas of openings A and their average

Table 2.2 Material data for fire-reduction

	k	T_1	T_2	T_8	T_{64}
Siliceous concreteHOT	0.00	15,000	800	570	100,000
Siliceous concrete COLD	0.00	3500	600	480	680
Main group concreteHOT	0.00	100,000	1080	690	1000
Main group concrete COLD	0.00	10,000	780	490	100,000
Light aggregate concreteHOT	0.00	100,000	1100	800	940
Light aggregate concrete COLD	0.00	4000	650	830	930
Russian Fire resistant concrete on Chamotte HOT + COLD	0.00	100,000	100,000	1150	1150
Danish Fire resistant concrete on Mo-Clay HOT + COLD	0.00	10,000	4000	710	1100
Hot rolled bars 0.2% stressHOT	0.000	6000	620	565	1100
Hot rolled bars 2.0% stressHOT	0.000	100,000	100,000	593	100,000
Hot rolled bars 0.2% stress COLD	1.000	100,000	100,000	100,000	100,000
Hot rolled bars 2.0% stress COLD	1.000	100,000	100,000	100,000	100,000
Cold worked bars 0.2% stressHOT	0.000	100,000	900	555	100,000
Cold worked bars 2.0% stressHOT	0.000	100,000	5000	560	100,000
Cold worked bars 0.2% stress COLD	0.580	100,000	5000	590	730
Cold worked bars 2.0% stress COLD	0.520	100,000	1500	580	650
C-w prestressing steel 0.2% stressHOT	0.000	2000	360	430	100,000
C-w prestressing steel 2.0% stressHOT	0.000	100,000	490	450	100,000
C-w prestressing steel 0.2% stress COLD	0.200	100,000	750	550	650
Quenched and Tempered 1500 MPa 0.2% stressHOT	0.000	1100	100,000	430	100,000
Quenched and Tempered 1500 MPa 2.0% stressHOT	0.000	3000	1400	450	100,000
Quenched and Tempered 1500 MPa 0.2% stress COLD	0.213	100,000	10,000	590	660
Quenched and Tempered 1500 MPa 2.0% stress COLD	0.213	100,000	10,000	590	660
Quenched and Self-tempered 550 MPa 0.2% stressHOT	0.000	6000	1150	540	700
Quenched and Self-tempered 550 MPa 2.0% stressHOT	0.000	100,000	100,000	590	700
Quenched and Self-tempered 550 MPa 0.2% stress COLD	0.418	100,000	100,000	700	900
Quenched and Self-tempered 550 MPa 2.0% stress COLD	0.437	100,000	100,000	700	900

opening heights h as

$$O = \frac{A\sqrt{h}}{A_{tot}}$$

The cross-section can be insulated for example by an insulation material or by a light concrete. This means that you can also apply the program for fire safety design of a composite structure or a super-light structure as presented in Chap. 3.

For CO_2 emissions, please consult Chap. 11 on Sustainability.

2.1.2 High-Strength Concrete

Some concrete codes like the Eurocode [3, 10] define as mentioned in Sect. 2.1.1, a strength of approximately 60 MPa as an upper limit for their application.

Researchers have for many years tried to develop concrete materials of a substantially higher strength by filling out the space between the cement grains with fine particles. The main hindrance they had for achieving that was electrical charges hindering them to place the particles that close.

However, about 1980 Hans Henrik Bache succeeded in doing so using the newly developed super-plasticizing additives [11]. These additives neutralize the electrical charges and make the concrete easier to cast and, as Bache realized, opened a possibility for making a more compact material. By means of this new technology, he could fill out the cavities between the cement grains with silica fume or other small particles.

He was able to give his new concrete a compressive strength of 240 MPa and he called it "Densit". He applied burned bauxite as an aggregate since the strength now became so high that the strength of ordinary granite would be a limiting problem. The colour was black, because carbon adhered to the fly ash.

The first author Hertz found the new strong materials promising for design of light structures with long spans and small cross-sections. He therefore asked Bache to send a pallet of test cylinders, so that he could investigate the properties of the new material and find if there were any limitations for the application of it.

The strong and stiff material gave rise to violent ruptures splitting the cylindrical specimens vertically when testing their compressive strengths. Hertz had foreseen that and provided the test machine with a safety screen. However, what he had not foreseen was that the material exploded when being slowly heated up by only one degree Celsius per minute in an oven in order to determine the residual compressive strength after heating and cooling.

The slow heating rate was chosen in order to avoid destructive thermal stresses in the specimens, but the test cylinders exploded at only 300 °C. Figure 2.3 shows an explosive fraction of one of the first cylinders tested.

It was immediately obvious, that the steam pressure trapped in the extremely dense material caused the explosion. [12–17].

Fig. 2.3 Exploded cylinder of Densit. *Photo* KD Hertz

The author also dried test specimens out at 100 °C and checked by weight that they did not contain any free water. These specimens did also explode proving that crystal water is sufficient for explosive spalling of such a dense concrete.

Explosive spalling appears to be a paramount hindrance for application of high strength concrete in buildings, where fire exposure is relevant. It has therefore been a subject for further investigations leading to the following limits for application of concrete in building design as given by [7]:

- Dry light-aggregate concrete with open pores is not susceptible to explosive spalling.
- Ordinary concrete of strength less than 80 MPa may suffer from spalling of corners or sides of fresh or wet structures within the first 20 min of a standard fire.
- High strength concrete stronger than 80 MPa subjected to compressive stresses or hindered thermal expansion is susceptible for explosive spalling.
- Free unloaded structures of ultra-high strength concrete of more than 150 MPa may spall explosively.

This leads to the following more general rule for application of concrete, where fire-exposure is relevant

> A concrete appears to be safe concerning explosive spalling, if its compressive strength is less than 60 MPa and its moisture content is less than 3% by weight.
> If a higher strength is applied, the temperature should not exceed 350 °C.

Table 2.3 High-strength concrete data

Compressive strength f_{cc}	100	150	240	300	400	MPa
Tensile strength f_{ct}	4.0	8.0	10	11	12	MPa
Initial E-modulus E_{c0}	43	45	50	55	60	GPa
Density ρ_c	2350	2400	2500	2550	2700	kg/m^3

By embedding the high-strength concrete into a light and insulating concrete, you may fulfil the conditions above.

Super-light and composite structures dealt with in this book therefore represent a possibility for utilization of the benefits of high-strength concrete that elsewise cannot be applied.

This can lead to lighter structures and longer spans etc. than whatcan be obtained with ordinary concrete structures.

High-strength concrete is developed towards still higher strengths. Since Bache's Densite of 240 MPa, strengths of 300$-$, 400$-$, and even 800 MPa have been achieved.

Table 2.3 shows examples of properties of high-strength concrete, which are known to be applied. They are based on 5% quantiles. Considerable deviations may occur depending on the actual recipes, aggregates, choice of small particles, additives, and the procedure applied when producing the concrete.

For some applications of high-strength concrete in super-light elements dealt with in this book, we apply a material with $f_{cc} = 110$ MPa, $f_{ct} = 5.0$ MPa, $E_{c0} = 43$ GPa, density 2360 kg/m^3, thermal elongation coefficient $\alpha = 1.1 \cdot 10^{-5}/°C$, heat capacity $c_p = 1.00$ kJ/m^3, and conductivity $\lambda = 0.90$ W/m°C.

For other applications, where the denseness is of importance as for example a strong outer surface layer of a tunnel, a floating building, or ships dealt with in Sects. 9.5 and 9.6, we consider high-strength concretes at the level of the original Densit.

It could be one with $f_{cc} = 300$ MPa, $f_{ct} = 11$ MPa, $E_{c0} = 55$ GPa, density 2550 kg/m^3, thermal elongation coefficient $\alpha = 1.1 \cdot 10^{-5}$ /°C, heat capacity $c_p = 1.00$ kJ/m^3, and conductivity $\lambda = 1.00$ W/m°C.

2.2 Light Concrete

2.2.1 Light Aggregate Concrete

As mentioned in Sect. 2.1.1, you can reduce the density of concrete by application of light aggregates.

Natural pumice from volcanoes is still widely applied, as it was in antiquity (see Sect. 1.2.3).

It may be transported long distances by ship. In Denmark, light aggregate concrete producers apply pumice from Iceland.

Other alternatives are expanded clay or expanded perlite. The quality of these aggregates varies with the raw materials and the processes applied by the producer.

Figure 2.4 shows a cylinder of a light aggregate concrete of density 700 kg/m^3 commonly applied for super-light structures (see Chap. 3).

The aggregate is expanded clay and it has a compressive strength of 3.5 MPa and a tensile strength of 0.30 MPa, initial E-modulus 3.5 GPa, thermal elongation coefficient $\alpha = 1.1 \cdot 10^{-5}$ /°C, heat capacity $c_p = 1.00$ kJ/m^3, and conductivity $\lambda = 0.35$ W/m°C.

As you can see, the concrete is quite porous, which means that there is no risk of explosive spalling, and that you can apply the material for sound damping surfaces as explained in Chap. 4.

Figure 2.5 shows ovens applied for production of expanded clay aggregates by a Danish factory. As explained in 10.2.2, the factory has now replaced 80–90% of the coal applied for heating the ovens with CO_2 neutral waste. The CO_2 impact is therefore halved from 0.27 to 0.14 kg CO_2 per kg of the resulting light-aggregate concrete.

Table 2.4 shows some possible material properties based on 5% quantiles. Light-aggregate concretes represent considerable deviations mainly due to different qualities of the aggregates (Fig. 2.4).

Fig. 2.4 Light aggregate concrete. *Photo* KD Hertz

Fig. 2.5 Expanding clay aggregate ovens. *Photo* KD Hertz

Table 2.4 Light-aggregate concrete data

Density ρ_c	600	700	900	1200	1800	kg/m^3
Compressive strength f_{cc}	3.00	3.5	5.0	7.5	15.0	MPa
Tensile strength f_{ct}	0.20	0.30	0.8	1.8	2.5	MPa
Initial E-modulus E_{c0}	3.00	3.5	4.5	7.0	16.5	GPa
Conductivity λ	0.30	0.35	0.40	0.45	0.90	W/m°C

2.2.2 Foam Concrete

Foam concrete has proven to be an applicable material in remote areas, where it may be difficult to get light aggregates. However, the recent development unveils several promising possibilities for the material.

You simply create a foam and mix it gently into a cement slurry. Even barber foam can be applied, but more robust foam liquids are developed for the purpose. Some companies like Aercrete produce transportable foam concrete mixing machines.

Table 2.5 shows some examples of foam concrete data from [18, 19].

Foam concrete has a fire-resistance equal to the one of light-aggregate concrete (Fig. 2.6). You can therefore estimate fire safety for structures with foam concrete using values for light-aggregate concrete in Table 2.2.

In practise, densities of above about 600 kg/m^3 are applied for load-bearing structures. However, at the author's university a number of projects have developed much lighter foam concretes. Such foam concrete is not for structural applications, but created as an insulating material, which can be cast in place and fill out all kinds of cavities.

Table 2.5 Light foam concrete data

Density ρ_c	100	300	600	700	900	1200	1800	kg/m³
Compressive strength f_{cc}	-	1.00	3.00	3.5	5.0	7.5	15.0	MPa
Tensile strength f_{ct}	–	0.05	0.20	0.30	0.8	1.8	2.5	MPa
Initial E-modulus E_{c0}	–	1.00	3.00	3.5	7.0	7.0	16.5	GPa
Conductivity λ	0.035	0.05	0.11	0.12	0.20	0.32	0.85	W/m°C

Fig. 2.6 Foam concrete. *Photo* KD Hertz

Jensen and Vahlgren [19] saw this potential and developed insulating foam concretes down to 200 kg/m³. Later, others have followed, at qualities with a density of 100 kg/m³ and conductivities comparable with that of mineral wool insulation have been obtained.

2.2.3 Aerated Concrete

Aerated concrete is widely applied for industrially produced building blocks and wall elements. The factory adds an aluminium powder to cement slurry and fill it into a train of small wagons on a track.

The slurry expands due to formation of bubbles and after some hardening, the volumes at the wagons are cut into blocks of the size wanted.

The wagons are then driven into an autoclave, which can be a tube of length 100 m in which the material is subjected to heat and pressure for some time. This process stabilizes the aerated concrete.

A typical aerated concrete has a density of 600 kg/m^3, and it has other physical properties comparable to a foam concrete or a light-aggregate concrete of the same density.

2.2.4   Pervious Concrete

You can design a concrete, which has a system of connected cavities between the aggregates with a void content typically close to 20%. Such concrete can drain water efficiently and hinder damages from repeating frost and thaw.

These materials are widely applied for roads in USA, and Iowa State University is a centre for development of pervious concrete. In cooperation with that university a PhD project made by Mia Lund [20, 21] developed a pervious concrete, which could be cast out in a height of more than 2 m, so that you can apply it as a draining and shear transferring material between arch and top slab in pearl-chain sandwich arch bridges. See 8.3.2. This material was used and tested in practise in the bridge shown in Fig. 8.17.

Typical material properties obtained for pervious concrete are a compressive strength of 10 MPa, E-modulus 16 GPa, Tensile strength 1.8 MPa, and a cavity content of 17% leading to a density of about 1900 kg/m^3.

2.3   Reinforcement

2.3.1   Slack Steel Bars

A steel material consists of grains with steel crystals. In the crystals, you find some impurities called dislocations, where a crystal line does not proceed all way through the lattice in the grain. When you load a bar of mild steel, the crystals at first deform elastically and the deformation disappears, when you unload the bar. This elastic deformation gives a stress σ that is proportional to the strain ε as $\sigma = E_{s0}\,\varepsilon$, where E_{s0} is the initial modulus of elasticity.

At a certain stress f_s the dislocations in the crystal move, and the steel does not regain its original size, when you unload it. The stress–strain curve will be as shown to the left in Fig. 2.7. We call this well-defined process yielding and we call this stress the yield stress. If the steel is mild steel, the strain is about 0.1% and the yield stress is about 240 MPa. The stress will fluctuate at this level for a while if you increase the strain. During this process, you create more dislocations in the crystal.

At a certain strain, the stress increases again, because you have created so many dislocations that they hinder each other's movement in the steel crystal. We call this phenomenon cold hardening or strain hardening. If you unload the bar, it will have a permanent deformation, and if you load it again, this cold-hardened steel will still

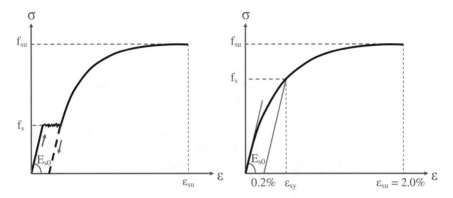

Fig. 2.7 Stress–strain curve for mild steel (left) and hardened reinforcing steel (right)

have the many new dislocations and it will not have a clear yield point. Instead, it will have a stress–strain curve as the one shown for hardened steel to the right in Fig. 2.7.

For such hardened steel, you define the "yield stress" f_s as the 0.2% stress by drawing a straight line with the inclination E_{s0} from the point 0.2% at the strain axis to the point, where it intersects the stress–strain curve of the cold hardened steel. From this point, you can consider the steel to deform irreversibly, and you apply the 0.2% stress as a yield stress for design of most steel structures. You can continue to increase the stress beyond that level contributing to a further cold hardening until the steel fails and you have reached the ultimate stress f_{su}.

Slack reinforcing bars were at first plain bars of mild steel with a yield strength of about 240 MPa and a modulus of elasticity $E_{c0} = 210$ GPa.

However, soon designers of reinforced concrete structures learned to apply cold hardened bars for the main forces with yield strength or 0.2% strength of about 550 MPa and the same initial modulus of elasticity. Often, the deformation was a combination of tension and torsion with the purpose of creating more dislocations. They also provided these bars with a corrugation to ensure a better grip between steel and concrete. We now only use mild steel for secondary reinforcement.

Instead of making cold hardened bars, industry began to improve strength of corrugated or deformed reinforcing bars chemically by adding carbon and later other material atoms to the crystal, which can anchor the dislocations. Deformed bars with carbon atoms have yield strengths of about 420 MPa. Today, you mainly apply slack deformed reinforcing bars with 0.2% yield strength of 550 MPa and an initial modulus of elasticity as $E_{c0} = 200$ GPa. This quality is mainly applied for deformed slack reinforcing bars in this book.

When you heat up a steel bar, the yield strength or 0.2% strength decreases to about the half at 500 °C because you need less mechanical stress to move the dislocations in the crystals [22]. For the same reason, you can create more dislocations by tensioning the steel further, and within the first 400 °C you can experience an increase of the cold hardened strength at large strains.

This is the reason why you can apply a smaller strength reduction for heated steel subjected to at least 2.0% strain than the reduction of the yield strength or 0.2% strength.

Table 2.2 therefore distinguishes between the reduction of 0.2% strength or yield, which should be applied for most steel structures and slack reinforcement. You can only apply the reduction of the 2.0% strength, if you can prove that you have a strain of at least 2.0%. [7].

You can e.g. find such strain of more than 2.0% for plastic design of prestressed reinforcement and of slack reinforcement in T-shaped beams, where the compression zone has a small depth compared to the internal lever arm to the reinforcement in the tension zone.

You can never find a strain of 2.0% in columns and usually not in ordinary slack reinforced beams or decks, because the deflections should then be so large that the components would fail due to them. Steel does not become more porous when heated as concrete, and the stress–strain curves are reduced by an affinity in the strain axis. This means that

$$f_s(T) = \xi_s(T)f_{s20} \quad , \quad E_{s0}(T) = \xi_s(T)E_{s020} \quad \text{and} \quad \varepsilon_{su}(T) = \varepsilon_{su20} > 2.0\%$$

Any point $(\varepsilon_{s20}, \sigma_{s20})$ at the stress–strain curve is transferred to $(\varepsilon_{s20}, \xi_s(T) \cdot \sigma_{s20})$.

2.3.2 Prestressing Steel

You mainly produce steel for prestressing strands by cold hardening, where large tensile strains provide it with a high ultimate stress. Due to the high stresses, you often apply 0.1% yield strength instead of 0.2% strength, so that you apply more acceptable strains and deformations for plastic design.

For most prestressed structures in this book, we apply prestressing steel with 0.1% yield strength of 1634 MPa, ultimate tensile strength of 1860 MPa, and initial E-modulus of 195 GPa. For pretensioned SL-slabs and other minor pretensioned structures, we apply single strands (of seven wires) of a diameter of 12.5 mm with a nominal steel area of 93 mm².

They are often pretensioned to about 72% of the 0.1% yield strength or 1176 MPa.

For larger pretensioned structures and for post-tensioned structures, we mainly apply strands of 16 mm with a nominal steel area of 150 mm².

Usually, you fix strands by wedges in the ends of a prestressing bed.

Factories anchor the straight pretensioned strands with wedges to large steel plates on heavy concrete blocks. The blocks are placed at each end of the prestressing bed, which is a mould of often about 100 m in length.

For in-situ cast structures or curved structures, you apply post-tensioned reinforcement.

Post-tensioning consist of cables positioned through an embedded post-tensioning duct. The duct can follow almost any desired shape. A cable consists of a number of strands, and the cable duct is often grouted with a mortar after tensioning the cable. At the ends, the cable is anchored to the concrete. Here, you cast special anchor blocks and reinforcement into the structure, to fix wedges for the number of strands in the cable. Usually, you place a splitting reinforcement at the end of the structure as well distributing the concentrated load of the anchorage to the full cross-section.

For straight prestressing steels, you can alternatively apply solid rods. They are provided with screw threads and nuts in the ends and may typically have a nominal diameter between 20 and 75 mm, ultimate tensile strength 1030 MPa, minimum 0.1% proof stress 835 MPa, and modulus of elasticity 185 GPa as seen for example in [23].

2.3.3 Slack Carbon Fibre Reinforcement

Designers and contractors apply still more carbon fibre reinforcement in new structures, and for repair of existing structures, where fire load is not important such as bridges and some industrial buildings.

Carbon fibre reinforcement has very high tensile strength and a modulus of elasticity comparable to steel. The density is only about 20% of that of steel.

Carbon fibre reinforcement does not suffer from rust as steel reinforcement does.

Designers can therefore apply it with the benefit of smaller or no cover thickness in many structures. An example of the use of carbon fibre in concrete hinges is shown in Sect. 9.3.2.

Usually, producers place a number of carbon fibres in a reinforcing bar or a carbon fibre texture in a flat band. In both cases, they embed the carbon fibres in a polymer. This is the main reason for the well-known lack of fire resistance. At the author's lab, we have made several fire tests of carbon fibre reinforcement and concrete structures with that reinforcement for example in a project made by Beck and Tvermose [24].

When we heat a carbon fibre bar, the polymer melts and evaporates at temperatures above approximately 200 °C. We can best describe the result as a horsetail of single carbon fibres, which seem to have a better fire-resistance than the polymer. We therefore see a potential for developing carbon fibre reinforcement with a better fire resistance.

Until that is developed, we have to protect the carbon fibre reinforcement from fire.

Here, designers can benefit from the application of light and well insulating concrete in the super-light structures keeping the temperature of the reinforcement of load-bearing parts below temperatures, where carbon fibre reinforcement melts (200 °C) or where high-strength concrete explodes (350 °C, see Sect. 2.1.2). This opens possibilities of new applications of both materials for creating very light structures with large spans and small impact on the climate.

Typical material values for carbon reinforcing bars are tensile strengths of 2000 MPa or 3000 MPa, with initial E-moduli of 250 GPa or 300 GPa, respectively. The ultimate strains are about 1% and no yielding occurs, so the material must be considered more brittle than steel reinforcement.

2.3.4 Prestressed Carbon Fibre Reinforcement

Prestressing of carbon fibre reinforcement would open new possibilities for design of super-light structures as described in Chap. 3 and pearl-chain structures described in Chap. 7.

As mentioned in Sect. 2.3.3, embedding the carbon fibres in light concrete may solve the problem of strength loss in case of fire. However, anchorage of prestressed carbon fibre reinforcement bars was a major problem in the past. However, research by Jacob Wittrup Schmidt has solved the problem [25, 26].

He applies a new design of an aluminium wedge of approximately double length of that steel wedge used for anchoring of prestressed steel strands. The system has been applied for external prestressed reinforcement for repair of a number of highway bridges, and it has been tested in several other structures. This means that the new technology of prestressed carbon fibre reinforcement may be considered applicable in general.

References

1. Desideri P, Nervi PJ Jr, Positano G (1979) Pier Luigi Nervi—A cura di Paolo Desideri, Pier Luigi Nervi jr, Giuseppe Positano. (In Italian) Zanichelli Editore, Bologna, 215p
2. Olmo C, Chiorino C (2010) Pier Luigi Nervi, Architecture as Challenge. Silvana Editoriale, Milan, 240p
3. CEN (2008) EN 1992-1-1 Eurocode 2, design of concrete structures, part 1-1. General rules and rules for buildings. Brussels. 225p
4. DS Danish Standards (1984) DS411, 3rd edn. NP-169-N Code for concrete structures. Danish Society of Engineers Copenhagen. 98p
5. Ritter W (1899) Die Bauweise Hennebique. (In German). Schweizerische Bauzeitung vol 33, Heft 7, pp 59–61
6. Hertz KD (2005) Concrete strength for fire safety design. Mag Concr Res 57(8):445–453
7. Hertz KD (2019) Design of fire-resistant concrete structures. ICE Publishing, Thomas Telford Ltd. ISBN: 9780727764447. London, 254p
8. DTU-Byg (2019) Software. Department of Civil Engineering, Technical University of Denmark. https://www.byg.dtu.dk/english/research/publications/software
9. ISO (1999) ISO 834-1 Fire-resistance tests—elements of building construction—part 1: general requirements. International Organization for Standardization, Geneva
10. CEN (2006) EN 1992–1–2 Eurocode 2, Design of Concrete Structures, Part 1–2. General rules – Structural fire design. Brussels. 97p
11. Bache HH (1981) Densified cement/ultra-fine particle-based materials, Aalborg Portland. Presented at the second international conference on superplasticizers in concrete, Ottawa

12. Hertz KD (1982) Eksplosion og reststyrke af varmepåvirket silikabeton (Explosion and Residual Strength of Fire Exposed Silica Fume Concrete. In Danish.) Report 162. Institute of Building Design (now Department of Buildings and Energy). Technical University of Denmark, 12 p

13. Hertz KD (1984) Heat-Induced explosion of dense concretes. Report No.166. Institute of Building Design, Technical University of Denmark. Lyngby. Presented at the CIB W14 conference in London May 1984) CIB W14/84/33 (DK), 20p

14. Hertz KD (1985) Explosion of Silica-fume Concrete. Fire Safety J 8(1):77

15. Hertz KD (1991) Silica fume concretes at elevated temperatures. key note speech at a special session on the author's findings at the ACI Spring Convention in Boston.

16. Hertz KD (1992) Danish investigations on silica fume concretes at elevated temperatures. ACI Mater J 89(4):345–347

17. Hertz KD (2003) Limits of spalling of fire-exposed concrete. Fire Safety J 38(2): 103–116

18. Aercrete (2009) Aercrete systemet. Aercrete Sweden

19. Jensen RL, Vahlgren A (2011) Insulation of super-light structures in Arctic regions. Final project in Arctic Technology, Technical University of Denmark, In Danish (Isolering af superlette konstruktioner i Arktis), 64p

20. Lund MSM, Hansen KK, Hertz KD (2016) Experimental investigation of different fill materials in arch bridges with particular focus on pearl-chain bridges. Constr Build Mater 124:922–936

21. Lund MSM (2016) Durability of materials in pearl-chain bridges. PhD thesis, Report R-341 Department of Civil Engineering, Technical University of Denmark. 157p

22. Hertz KD (2004) Reinforcement data for fire safety design. Mag Conc Res 56(8):453–459

23. Macalloy (2009) Macalloy 1030 Post tensioning system. Macalloy Ltd. Sheffield. 8p

24. Beck PP, Tvermose HH (2004) Fire exposed carbon fibre bands. In Danish (Brandpåvirkede kulfiberbånd) MSc project, Technical University of Denmark, 120p

25. Schmidt JW (2011) External strengthening of structures with prestressed CFRP tendons. PhD Thesis, COWI Ltd and Technical University of Denmark. 289p

26. Schmidt JW, Krabbe Sørensen JNO, Hertz KD, Goltermann P, Sas PG (2017) CFRP strengthening of RC beams using a ductile anchorage system. In: Proceedings of the eighth international conference on fibre-reinforced polymer (FRP) composites in civil engineering, pp 344–349

Chapter 3
Super-Light Structures

Abstract The core of Super-light technology is to apply the correct and necessary materials in the optimal position to create a minimal structure. The chapter deals with how to utilize knowledge of concrete types in guiding the forces through a construction. A Direct Engineering approach is introduced to learn how it is possible to choose where the forces should be.

3.1 Super-Light Theory

3.1.1 Minimal Structures

Usually, engineers would like to design structures for a minimum of material consumption, cost, construction time, CO_2 emission, pollution, and other disadvantages. Figure 3.1 shows an example. It is one of the first super-light building projects from 2010 designed by Bjarke Ingels Group (BIG) Architects with Werner Sobek as engineer intended for a Building 324 at the Technical University of Denmark.

Builders and researchers try in general to investigate and develop optimal structures like the vault (Figs. 1.7, 1.8, and 1.9), the cupola (Figs. 1.13 and 1.14), the arch bridge (Fig. 1.10), or the suspension bridge as seen at the top of Fig. 3.2.

Suspension bridges are often referred to as minimal structures, because the forces follow a path that results in a minimum use of material in places, where the amount of material and its weight matters as for example at the free span of the bridge.

The suspension bridge has tension cables in an optimal shape, guiding the forces to two towers in compression. However, you should also design the bridge to resist the horizontal tension forces, which often give rise to some large, expensive and material consuming foundations.

A theoretical alternative would be to counterbalance the horizontal tension forces at the ends of the cables with compression in the bridge deck. This is almost never possible in practise, since the deck is usually designed as a light structure that is too slender to be able to resist these large compression forces because of the risk of buckling.

Fig. 3.1 Project for a super-light structure 2010. (BIG Architects)

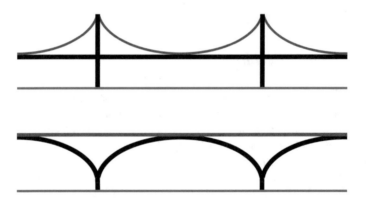

Fig. 3.2 Examples of minimal structures

At the bottom of Fig. 3.2, you can see an arch bridge, which is the counterpart of the suspension bridge, where you have exchanged compression and tension in all structural members. Here, the curved arches have optimal shapes to lead the load of the bridge as compression forces to the foundations. If you only apply a single arch, the compression arch may also need some large, expensive and material consuming foundations to resist the horizontal lateral part of the arch force.

However, if you supplement the arch with two half arches as shown in the figure, you get the full counterpart of the suspension bridge, and now the bridge deck can resist the horizontal forces, because these forces are now in tension and therefore, they do not give problems with buckling. We consider this type of arch structure and its application in more details in Sect. 7.2.3.

The schematic drawings on Fig. 3.2 do not show all details of importance. Especially, they lack structural members transferring the continuous weight of the deck to the bearing cables of the suspension bridge and to the arches of the arch bridge. These members are hangers in tension at the suspension bridge. For the arch bridge, they may be columns in compression as shown in Fig. 7.19 or even a lattice as in Fig. 7.12.

Designers like Gustave Eiffel (Fig. 7.8) and Edgar Cadoso (Fig. 7.19) have developed and refined minimal structures for their large bridges, and researchers like [1–3] have developed general theories for minimal structures. However, when you reduce the size of the structure, you will find it more difficult to apply a minimal shape for a design that is optimal with respect to all parameters.

The smaller you make the structure, the closer, you have to place the hangers or columns to benefit from the curved shape. The connections then occupy a relatively larger part of the design and of the total resource consumption. The design therefore becomes more clumsy and less optimal.

In buildings, most dimensions are considerably smaller than in the large bridges, and it is even more clear that an optimal design requires a holistic approach. An optimized design depends on fulfilment of a number of very different functional requirements like sound insulation, fire resistance, sound damping, integrity, heat insulation in addition to the load-bearing capacity. The shape of an optimal structure can therefore seldom be the same as the one of a construction, which is minimal with respect to material consumption for the load-bearing function alone. A few gifted master builders like Appollodorus from Damascus (Fig. 1.12) and Piere Luigi Nervi, see Figs. 3.3 and 2.1, have shown us examples where a building shape coincides with a minimal structure [4, 5].

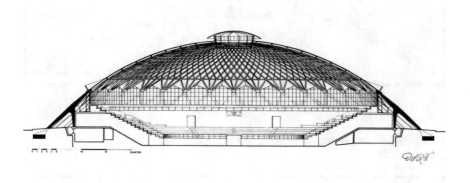

Fig. 3.3 Palazzetto dello Sport by PL Nervi 1958. (Drawing OSR Frederiksen)

3.1.2   Super-Light Principle

In 2007, the first author invented super-light structures in order to solve the problem of an optimal outer shape, that would not necessary be identical with the minimal shape of the load-bearing structure [6]. You achieve this by designing a primary minimal load-bearing construction, and place it in a light structure with an outer shape fulfilling a number of other requirements. The outer shape could for example be a rectangular beam that provides you with a horizontal surface on top. The light material, which in this case could be light concrete, transfers the load from the top of the beam to the embedded primary load-bearing construction, which in this case could be a curved arch of strong concrete. Furthermore, the light concrete stabilizes the stronger arch against buckling, so that you can reduce the required size of the arch cross-section and reduce the amount of strong concrete applied.

You may for example design a primary load-bearing structure as an arch (Fig. 3.4) of an ordinary 55 MPa concrete of density 2300 kg/m^3 and cast an outer structure of light aggregate concrete of density 600 kg/m^3 with a compressive strength of only 3 MPa. The strong arch may require a curved and expensive mould like those, we applied in the very first super-light test structures as for example the double beam shown in Fig. 7.10 made by Larsen [7]. The expensive mould has been a main reason why builders have rarely applied arches and vaults for more than 50 years in most industrialized countries.

We therefore soon found a more economical solution as shown in Fig. 3.5, where we hang a tube with the shape of the arch upside down, cast the light concrete around

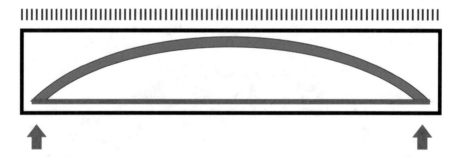

Fig. 3.4 Super-light principle with a strong arch in a light material

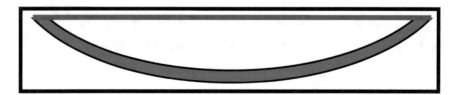

Fig. 3.5 Concrete arch cast in a tube and embedded in light concrete

it, and then cast the strong concrete in the tube. The tube could be a duct similar to those applied for post-tension cables, but you may apply less expensive solutions, because the light concrete gives a smaller mould pressure. Later, we invented the pearl-chain principle [8] as an inexpensive alternative to the curved mould. Chapter 6 explains more about this.

3.1.3 Benefits and Goals

The super-light principle allows you to design a structure with a minimum of material consumption, CO_2 emission, and transport costs and still fulfill the many different functional requirements.

The application of a composite structure with two different concrete materials with different densities and elastic moduli gives you another benefit. The concrete parts oscillate with different eigenfrequencies and the structure therefore reduces sound energy to heat.

This means that you can obtain a better sound insulation than possible, if you only applied one material. We utilize this in SL-decks (Chap. 4) and composite walls (Chap. 5) to reduce material consumption and CO_2 emission of structures for domestic buildings and other places, where sound insulation is required or wanted (Chap. 11).

The light concrete is very fire resistant [9] and has an insulating effect. This means that if you embed strong concrete in an outer shape of light concrete, you protect the strong load-bearing concrete from fire and you may obtain a considerably increased fire-resistance. We explain this further for the super-light deck elements in Chap. 4.

You may even design your structure so that you keep the temperature of the strong concrete below 350 °C. This means that you can apply a high-strength concrete of more than 80 MPa to carry the load without a risk of explosive spalling [9].

Explosive spalling has so far been a hindrance for application of high-strength concrete in buildings and tunnels. However, super-light structures solve this major problem.

Since light concrete can also stabilize the strong, you can now apply very slender cross-sections of high-strength concrete. This allows you to design super-light structures of extremely low weight and for example obtain long span widths.

3.2 Direct Engineering

3.2.1 Direct Engineering Principle

In traditional structural engineering, you assess the outer loads and choose a shape of the structure according to your experience of how structures usually are.

Then, the next task is to find out, how the forces are distributed in the structure. To do this, you make a structural analysis.

This is a rather demanding discipline, where you make qualified presumptions concerning stress–strain relationship of involved materials, how stresses can be distributed, how plastic hinges can occur, how cracks may develop and change the stress distributions, etc.

Finite element programs are often applied to give an idea of how stresses and forces could be distributed, and the outputs of these programs are approximations depending on a number of assumptions. Finally, you find a design of the structure that can contain a distribution of forces, which can resist the loads.

The principle of super-light structures allows you to reverse the engineering process.

Here, you can decide, where you want the forces to be. This means that you decide an optimal shape of the load-bearing parts of strong concrete. These may for example be a minimal structure of arches as the simple example shown in Fig. 3.4 and perhaps you supplement them with some elements in tension.

Then you fill the outer shape with a light concrete that stabilises and protects the strong.

Since you know the forces that you want the single elements to transfer, you can immediately design the strong elements to resist them.

Since the strong elements have a considerably larger stiffness (about 10 times larger modulus of elasticity) than the light concrete, you can be sure that each force will find the place, where you have decided it to be. This means that you are not required to make a structural analysis.

Especially not, if you consider that you have a lower bound solution. (See Sect. 3.2.2).

Of course, you may do so in order to obtain an extra check of your structure.

This way of working with construction design appears to be straight forward, and we therefore call it "Direct engineering" in contrast to the traditional that appears to be backwards and indirect.

You may say that in direct engineering, you decide how the structure should perform instead of accepting that the structure determines how you should perform (Fig. 3.6).

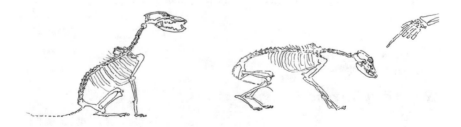

Fig. 3.6 Direct engineering

When you design super-light structures with direct engineering, it is sometimes helpful to consider the structure as a skeleton of strong concrete embedded in a body of soft parts of light concrete perhaps interacting with sinews of tension members such as reinforcement.

The analogy to the body anatomy can inspire you to find new solutions for structural problems.

As mentioned, you can make the curved shapes of the compression zones inexpensively and effectively as pearl-chains according to Chap. 6. It is possible to design pearl-chains of mass-produced straight elements, which constitute the curved structures like spines in a skeleton analogy.

Tension members may be ordinary steel reinforcement bars and cables, but they may also be carbon fibre rods or lines. In many cases, this is a new possibility, because the light concrete can provide a fire-protection required for application of such materials. Carbon fibre does not corrode. This therefore opens new possibilities of placing tension reinforcement in light concrete exposed to a wet climate or reducing concrete cover layers.

You may also design tension members as prestressed concrete elements. They can be mass-produced as pre-tensioned structures by an element factory. They can also be designed as post-tensioned concrete elements with a cable duct, and in this way, you can even create curved pearl-chain tension ties of prestressed concrete.

The benefit of these prestressed concrete structures is that they can resist tension as unloading of the pre-compression in the concrete. This has the advantage of a considerably larger axial stiffness, and consequently a smaller deflection and higher loads before cracking of the super-light structure.

Like carbon fibre reinforcement, you can apply prestressed concrete tension members embedded in a light concrete without problems of corrosion.

3.2.2   Lower Bound Solutions

If you can find a distribution of forces that can sustain the load on the structure, it is statically possible. If the structure can sustain the forces in any point, it is safe.

A statically possible and safe force distribution is per definition a lower bound solution.

If the internal force distribution in the physical structure is not as postulated, the structure is either able to sustain the applied load, or the forces may redistribute through cracking and plastic deformation into new positions for example as the postulated, where they can sustain the load. Since the postulated force distribution is proven to be possible and safe, the structure at least has one possible way of resisting the load, and it can therefore not collapse. This argumentation gives the third criterion that the internal force distribution should be able to change from the actual one in the physical structure, to the postulated one in the calculation model. This is often referred to as the structure having a sufficient rotational capacity.

No matter how you calculate a force distribution, you have to make simplifications and presumptions such as application of elastic-, plastic-, or other calculation models, a crack distribution model, a material model etc. This means that you actually always calculate a system that is different from the physical structure.

Therefore, if you find that your force distribution is safe, the calculation will be a lower bound solution. It is lower bound, because there may be another force distribution giving a larger resistance to the load. If that is the case, this other force distribution will be the one applied of the structure at the ultimate limit state. These considerations give the following conclusion:

At the ultimate limit the forces and stresses in a structure will be distributed so, that they give maximum resistance to the load, and that any other postulated safe and statically possible distribution will be a lower bound solution provided that the structure can redistribute forces from the postulated to the ultimate.

The principle of lower bound solutions has led to simple design methods such as the strip method introduced by Hillerborg [10], where a slab is considered to consist of two crossing sets of strips, where you postulate that the strips in one direction resist a certain part of the load, and de others the residual load.

Other researchers, as for example by Nielsen and Hoang [11], have developed theories about lower bound solutions further.

In super-light structures, you create a core or "skeleton" of strong concrete, where you postulate that the compression forces will be.

Perhaps you also have the skeleton interacting with some tension members and then you fill the outer shape with a light concrete material.

If the skeleton can resist the loads when the light concrete stabilizes it, you have a lower bound solution, and you have designed the structure.

References

1. Andersen S (1979) The accumulated energy consumption for building materials. Report 134, (in Danish) DTU Byg (Inst.of Building Design) 1979, 198p.
2. Andersen S (1980) The accumulated energy consumption for dwellings. Report 137, (in Danish) DTU Byg (Inst.of Building Design) 1980, 124p.
3. Reitzel E (1979) From fracture to shape. (in Danish), Polyteknisk Forlag, 269p
4. Desideri P, Nervi PJ Jr, Positano G (1979) Pier Luigi Nervi—A cura di Paolo Desideri, Pier Luigi Nervi jr, Giuseppe Positano. (In Italian) Zanichelli Editore, Bologna, 215p.
5. Olmo C, Chiorino C (2010) Pier Luigi Nervi, Architecture as Challenge. Silvana Editoriale, Milan, 240p
6. Hertz KD (2010) Light-weight load-bearing structures. (Super-light structures). Application no. EP 07388085.8. European Patent Office, Munich. November 2007. Application no 61/004278 US Patent and Trademark Office. November 2007. PCT (Patent Cooperation Treaty) Application PCT/EP2008/066013, 21. November 2008. Patent 2010.
7. Larsen F (2007) Light Concrete Structures. Department of Civil Engineering, Technical University of Denmark, MSc Project, p 117
8. Hertz KD (2012) Light-weight load-bearing structures reinforced by core elements made of segments and a method of casting such structures. (Pearl-Chain Reinforcement). Application no. EP 08160304.5. European Patent Office, Munich. July 2008. Application no 61/080445

US Patent and Trademark Office. July 2008. PCT (Patent Cooperation Treaty) Application PCT/EP2009/052987, 13. Mar 2009. Patent 2012.

9. Hertz KD (2019) Design of fire-resistant concrete structures. ICE Publishing, Thomas Telford Ltd. ISBN: 9780727764447. London, 254p

10. Hillerborg A (1975) Strip method of design. Cement and Concrete Association. 256p

11. Nielsen MP, Hoang LC (2016) Limit analy CRC Press, Taylor & Francis Group. First published in 1984:816p

Chapter 4
Slabs and Beams

Abstract The mass produced SL-Deck is a slab, which is based on the technology of Super-light concrete. It is light-weight, and has many benefits such as fire safety, sound insulation and flexibility regarding geometry and connections. The SL-Decks can be used in ordinary buildings, in beamless buildings, and for many other applications.

4.1 SL-Decks

4.1.1 Prototypes of Super-Light Deck Elements

When we invented the super-light structures described in Chap. 3, we could see a wealth of interesting applications, where builders could benefit from the new technology.

It was soon realized that producers of the first components would like to have a market of a certain size, where the products meant a considerable improvement in terms of a lower price for a better performance. Prefabricated deck-elements represented such a market.

In cooperation with BIG Architects, their applications were investigated in advanced building design [1]. By creating a mass-produced super-light deck element, material consumption and CO_2 emission could be reduced and fire-resistance and flexibility improved. Odd geometries of decks could be created with large holes, and embed services such as electricity, floor heating, water supply, and sewers in the elements.

In addition, we could prepare transfer of moments for vertical load in continuous slabs and shear for horizontal load to other structural components, where the slabs act as horizontal shear plates. Fixings could also be implemented for example for balconies and for preliminary handrails needed at the building site.

We therefore started to design a super-light deck element called the SL-deck.

Soon, it was realized that it should consist of some blocks of light-aggregate concrete, which could serve as permanent moulds for a stronger concrete.

K. D. Hertz and P. Halding., *Sustainable Light Concrete Structures*, Springer Tracts in Civil Engineering, https://doi.org/10.1007/978-3-030-80500-5_4

The strong concrete can carry the load in small arches across the blocks. The arches rest on reinforced ribs in two directions between the blocks. This means, that the strong concrete constitute a ribbed structure with vaulted plates between the ribs.

At that time in 2011, concrete factories were not able to cast strong- and light concrete together. We therefore asked a light aggregate concrete factory to assist us casting the first blocks. In cooperation with them, we found a shape for the blocks, which a machine could easily cast, and which could adhere to a strong concrete.

A series of blocks was made and transported to a concrete element factory that produced a number of prototype deck elements. Figure 4.1 shows the precast blocks placed on a 1.2 m wide and 100 m long track. The structure was reinforced with pre-tensioned 12.5 mm tendons along the track, and slack deformed (corrugated) bars across. Then we cast the strong concrete on top of this. Next day, we could lift the super-light deck elements off the track.

Comprehensive tests were made of the prototype elements (Fig. 4.2) for bending [2], shear [3], and anchorage [4]. In addition, we tested them for fire resistance. We considered the load case of fire very important, because new requirements for heat insulation have increased fire exposure representing a doubling of the required standard fire resistance [5]. See more about this subject in Sect. 4.1.4 and in Hertz [6].

Finally, a large effort was made for designing and testing the prototype elements for sound insulation in cooperation with the acoustical department of the consulting engineering company Grontmij [7]. Usually, the requirement of a proper sound insulation gives rise to design of heavy decks and walls and hereby a larger production of CO_2. See more about this subject in Sect.4.1.3.

Fig. 4.1 Blocks for 1.2 m wide prototype SL-decks in 2011. *Photo* KD Hertz

Fig. 4.2 Bending test of prototype SL-deck. *Photo* KD Hertz

The Danish Building Regulations has for example a minimum requirement of 55 dB sound insulation. This is a main hindrance for design of light building structures saving materials.

The Technical University of Denmark applied a number of the prototype deck-elements for balconies in a new building 324. The architect Christensen and Co. designed balconies as hung-down deck structures in two levels (Fig. 4.3). Ordinary

Fig. 4.3 Hung down balconies of prototype SL-decks. *Photo* KD Hertz

deck elements could not adopt the concentrated forces from the bearing rods, but SL-decks allow incorporation of such bearings. In addition, the balcony decks included blade connections so that one deck can rest in plane on the next without application of beams. (See more about that in Sect. 4.2.6).

4.1.2 Mass Produced SL-Decks

Based on the experience from design, test and application of the prototype decks, we designed the first SL-deck for mass production as shown in Fig. 4.4. The width was assessed to be 2.4 m, which is the double of that of the prototype and maximum of other prefabricated deck elements. We introduced a wider groove between the light-aggregate concrete blocks at the middle of the cross-section allowing implementation of special services or a duct for a post-tensioning cable. Figure 6.7 shows such elements, where the cable duct is visible in a recess, so that you can connect it to the duct of the next element before casting mortar in the joint.

This means that we have prepared the SL-deck for application as a inexpensive, mass-produced element for pearl-chain structures. The SL-deck is therefore not only intended to be used as slabs in buildings but also as components for arches and vaults and for bridge design as explained in Chaps. 6 and 7.

Figure 4.5 shows 100 m long and 2.4 m wide tracks for casting SL-decks, In the prestressing bed, magnetic mould steel plates separate the decks. The sides can turn and automatically de-mould the elements. We shaped the sides in the longitudinal joints as shown in Fig. 4.6 so that it can transfer horizontal and vertical shear to the next element. The profile allows pouring of mortar in the joint without a shutter

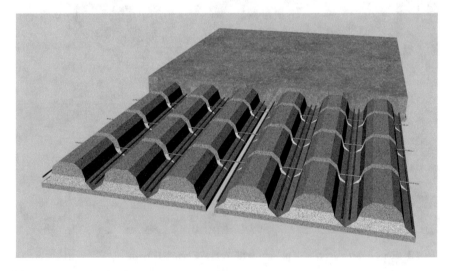

Fig. 4.4 2.4 m wide SL-decks

Fig. 4.5 2.4 m wide tracks for SL-deck production. *Photo* KD Hertz

Fig. 4.6 Detailing of the longitudinal sides of a SL-deck. *Photo* KD Hertz

board beneath. Special bearing knobs in the sides of the elements serve to lift the longitudinal reinforcing bar in the joint in order to embed it in the mortar ensuring its bond to the deck and to protect it against moisture and corrosion from the bottom of the joint.

At first, the factories usually place a thin layer of mortar on the mould, upon which the blocks are placed. This ensures a smooth underside of the elements, which customers usually want for their ceilings in domestic buildings.

Fig. 4.7 Automatic block casting machine for SL-decks. *Photo* KD Hertz

If wanted, the factory can keep the sides free from this bottom mortar layer to that there are voids along the element sides. This allows the contractor to fill the voids at the building site and thereby obtain a smooth unbroken ceiling without visible joints between the elements.

Figure 4.7 shows an automatic machine developed to cast the light-aggregate blocks. It can place blocks, where you have programmed it on a 100 m long track in two hours.

This means that you can omit blocks and hereby create spaces in the deck, where you can have strong concrete, holes, services, or inlayed structural elements such as bearings, beams etc.

Figure 4.11 shows an element with an inlayed beam at the end for lateral support.

Figure 6.17 shows an advanced example of a post-tensioned inlayed beam designed for bending, shear and torsion.

The strong concrete applied has usually a compressive strength of 55 MPa, and the density of the light-aggregate concrete is about 700 kg/m^3. The machine casting the light aggregate concrete blocks has a row of six moulds (for 2.4 m SL-deck widths).

The number of moulds used in each row and their position can be controlled. The machine vibrates the light-aggregate concrete, lifts the moulds, and proceeds to a new row. Then, the factory workers places tendons, which are usually 12.5 mm, along the track and pretension them. Then slack reinforcement, inserts and services etc. are positioned and finally, the strong self-compacting concrete is cast. The strong concrete penetrates the light before the light hardens and creates a zone of a couple of centimetres, where the quality gradually changes from the one to the other without forming a surface separating the two materials. Even at full-scale failure tests, the light blocks do not separate from the SL-deck.

By combining pre-programmed placing of blocks and placing of magnetic fixed mould pieces, you can create deck elements of any shape, you want. Figure 4.8 shows an irregular element lifted in place at a building site in Copenhagen Harbour.

Figure 4.9 shows a SL-deck element that demonstrates possibilities of making holes in the decks. The large rectangular holes were for skylights, and the engineer placed inlayed reinforced beams across the element between them. The element is still so freshly produced, that you can see the position of the light-aggregate blocks

Fig. 4.8 Irregular SL-deck element. *Photo* Abeo Ltd

Fig. 4.9 SL-deck with holes spanning over several bays. *Photo* KD Hertz

Fig. 4.10 Wedge shaped SL-deck elements for a round building. *Photo* Abeo Ltd.

as a pattern at the bottom caused by differences in moisture. The element is also an example of a continuous SL-deck designed to span over a number of rooms. This is obtained by placing top-reinforcement so that negative bending over the intermediate supports can be resisted.

By utilizing the option of continuous decks, you can make fewer crane lifts at the building site and transport fewer elements. You may also increase the possible span lengths by designing elements with fixed end supports.

Figure 4.10 shows how a series of equal wedge shaped SL-deck elements with rounded ends constitutes a floor structure of a round building. The picture also demonstrates how you can place preliminary safety handrails using inserts in the deck prepared for the purpose at the factory. This saved the workers 1–2 days of drilling and applying fixings for the safety handrails at the building site, which they would normally do for other deck solutions.

SL-decks are produced in thicknesses typically varying from 180 mm over 220 mm, 270 mm, 320 mm, and up. The strong concrete has typically a minimum thickness of 40 mm above the light blocks. Either you can choose to increase the block height as the total deck height increases, or you can keep the blocks height and just increase the total deck height. If you choose to increase the block height, it is done by varying the height of the inclined sides of the blocks, where all other measures of the block geometry remain constant.

The pretensioned tendons are usually 12.5 mm in diameter with a cross-sectional area of 93 mm^2. Two of them are placed at the sides of the element with 42 mm from the bottom to the centre line, and the others in the grooves between the blocks a levels of 60 and 92 mm from the bottom to the centre lines. 20 mm of this height from the bottom is insulating light-aggregate concrete.

Fig. 4.11 SL-deck element with inlayed beam for lateral support. *Photo* KD Hertz

Top reinforcement tendons are typically placed 30 mm from the top surface of the element to the centre lines.

At each end, you find a zone of massive concrete with a length of at least 200 mm, where the pretensioned reinforcement is anchored, and where the deck element is supported. This means that the presence of deck ends in joints between walls and decks does not introduce weak areas affecting the strength of a wall in a multi-storey building. As shown in Sect. 9.1, this may be more than sufficient for anchorage of the tendons because the compressive stresses from the bearing increases the splitting strength at maximum load.

As mentioned, the first Danish factory produces SL-decks of width 2.4 m. The cross-section of these elements has a maximum of six light-aggregate blocks of width 375 mm as shown in Fig. 4.12. This width is the maximum obtainable for other mass-produced concrete deck elements without a cross-reinforcement. However, SL-decks have a cross-reinforcement and they do not have this limit in order to avoid longitudinal rupture. The limit of the width is therefore solely determined by the width that can be transported by trucks without special permission. This is found to be 3.0 m, which has therefore become the new standard width for European factories. The cross-section of these elements has a maximum of eight light-concrete blocks each of width 350 mm as seen in Fig. 4.12 leaving two wider grooves, where ducts for post-tensioning cables can be placed for application in pearl-chains (Chap. 6).

In USA, two factories are built and more are on the drawing board.

Here the SL-deck is called T-slab, and the standard width is 12' or 3.6 m. This large width suites well to the American standards of building and allows very few expensive crane lifts.

SL-Deck 180, 220, 260 ... mm

2.4 m

3.0 m

3.6 m

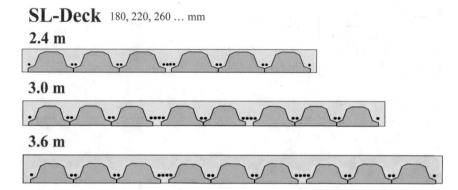

Fig. 4.12 Produced SL-deck cross-sections, 2,4 m and 3.0 m in Europe, 3.6 m in USA

4.1.3 Fire Resistance of SL-Decks

As mentioned in Sect. 4.1.1, we considered fire safety of deck constructions important.

Since 1990, building regulations prescribed improvement of building insulation in order to save energy. The result is, that a larger part of the heat from a fire remains inside, and that windows seldom break. This reduces ventilation and increases fire duration and temperatures in structures. You consider this, when you design a structure for a fully developed fire. However, you often make an alternative design for a fixed required fire-resistance time of a standard fire curve without a cooling phase according to ISO 834 [8].

The authorities have not yet adjusted these fixed requirements in the building regulations according to the better insulation of the buildings. If you for example compare a temperature of a reinforcing bar, you can see, that you should double the required standard fire-resistance time, if a standard fire design should still be relevant for the real fully developed fire in a modern building [6]. We therefore designed the SL-deck elements for a 240 min standard fire-resistance, so that it can withstand the same fires, as we previously represented by a 120 min requirement.

We first tested two prototype slabs before we made an accredited test on a mass-produced 220 mm thick, 2.4 m wide, and 6 m long SL-deck (Fig. 4.13).

After 240 min standard fire exposure, the bottom of the deck was 1154 °C, the prestressed reinforcement was 389 °C, and the top of the deck was 58 °C. The deck was unharmed and the thermal deflection disappeared, when the deck cooled down.

4.1.4 Sound Insulation of SL-Decks

Sound insulation means reduction of noise from one room to the next across a separating deck or wall. Deck structures are required to have a reasonable sound insulation

Fig. 4.13 SL-deck element after 240 min standard fire test. *Photo* KD Hertz

if users of a building should not disturb one another. In most countries, the building regulations require a minimum sound insulation, and in Denmark, it is 55 dB. Sound insulation requirements are major hindrances for making light sustainable structures, because you need mass in order to fulfil them. If a deck is made of massive concrete, you must have 440 kg mass per m^2 to obtain a 55 dB sound insulation.

A SL-deck consists of heavy and light concrete. The two materials oscillate differently, and the deck therefore reduces some of the noise to heat. This is the reason why, the SL-deck can obtain more than 55 dB sound insulation with a weight of only 350 kg/m^2.

We tested prototype decks [7] and made accredited tests of the mass-produced 220 mm SL-deck.

It documented that a bare deck element had a sound insulation of 57 dB, and installed with a standard floor, the sound insulation was 58 dB, where the requirement is 55 dB. (Fig. 4.14).

Step noise or impact sound level was measured with a standard impact machine to be 47 dB, which is 6 dB better than required.

4.1.5 Sound Damping of SL-Decks

Where sound insulation is reduction of noise transfer between rooms, sound damping or sound absorption is reduction of noise in the room itself. Bare concrete surfaces may reflect sound and give rise to a hard acoustic (strong reverberation) in buildings. Often, builders apply wood-wool claddings in order to damp the sound. The porous

Fig. 4.14 SL-deck element tested for sound insulation and step noise. *Photo* Abeo Ltd.

Table 4.1 Sound absorption of light-aggregate concrete compared to wood-wool

Frequency in Hz	125	250	500	1000	2000	4000
Light-aggregate	0.15	0.19	0.44	0.63	0.52	0.75
Wood-wool	0.08	0.20	0.45	0.80	0.66	0.85

surface of the wood-wool absorbs the sound energy by means of a kind of friction, when the air oscillates in the small cavities of the material.

When you apply SL-decks, you can have a similar sound damping by omitting the bottom layer of mortar, which the factory usually place in order to deliver a smooth surface.

If you omit that, you will have a free porous surface of the light-aggregate concrete that can provide a sound damping effect. Table 4.1 shows that the sound damping even as a function of frequency is almost identical to the one of wood-wool slabs. Figure 4.15 shows a classroom at Gammel Hellerup Highschool, where BIG Architects utilized the sound damping of the SL-decks in the ceiling.

4.2 Super-Light Deck Details

4.2.1 Fixed End Connections

If you design a continuous deck element supported by e.g. a number of walls, you can obtain longer span widths and save material and CO_2. Utilizing fixed ends is a

Fig. 4.15 Classroom with sound damping SL-decks. *Photo* KD Hertz

way to obtain longer spans and save CO_2 as well. Moreover, both design options save crane lifts and mounting time at the building site. SL-decks like the one shown in Fig. 4.9 are therefore often designed with several continuous spans, and provided with a top reinforcement in order to resist the negative moments at the supports.

If shear is a problem near a supporting wall, you may design a massive zone by omitting to place some blocks and reinforce it with stirrups as a beam.

You may also provide the deck element with holes to connect vertical wall reinforcement.

However, the length that a truck can carry in ordinary traffic and the weight that the applied cranes can lift, represent limits for the length of the deck elements. You may therefore wish to create a joint, where you can make a fixed end connection between two deck elements over a supporting wall. Figure 4.16 shows a simple design of such a joint. Slack top reinforcement is placed at the ends of the SL-decks, where you can expect negative moments and this slack reinforcement is connected to steel plates, which can be simply connected with bolts to the similar plates in the other element. The holes required for inserting the bolts are filled with mortar at the same time as the rest of the joints.

Thinking of the disassembly process of the building, a weak mortar can be used in the hole, so that the bolt can be exposed again for easy dismantling. (This is explained in more detail later in Sect. 9.2.1).

Alternatively, you can provide the deck elements with grooves, where the contractor at the building site can place a top reinforcement over the joint at the wall.

Fig. 4.16 Fixed end connection between SL-decks. *Photo* Abeo Ltd.

4.2.2 Connection to Balcony

Fixed end connections are also required for cantilevered balconies. Figure 4.17 shows an example, where top reinforcement and inclined shear reinforcement are concentrated in two prefabricated zones of low conductivity between a SL-deck and a balcony.

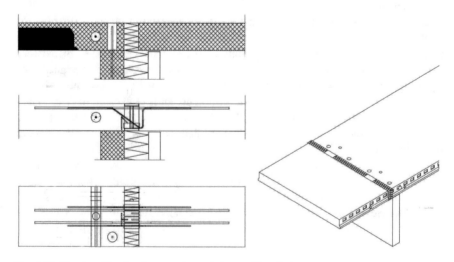

Fig. 4.17 Fixed end insulated connection to balcony. Abeo Ltd.

4.2.3 Connection to Column

If a column should support a SL-deck element without application of a beam, you can create a zone of strong concrete above and around the column by omitting blocks in the area. The zone of strong concrete may also serve to transfer forces from a column above the deck to the column below, and holes can be made in the deck in order to connect vertical reinforcement combining the column elements. The deck usually has to be supplied with horizontal reinforcement in two directions in order to resist negative moments. Probably, it will also be necessary to insert stirrups or inclined shear reinforcement in the deck around the column connection.

4.2.4 Safety Handrails

As shown in Fig. 4.10, the contractor can place safety handrails in fixings cast into the deck at the factory, and usually, the elements are lifted in place with the handrails mounted.

This saves time at the erection site for drilling fixings and mounting the handrails.

4.2.5 Services in SL-Decks

Sockets for vertical pipes and tubes for horizontal pipes and electrical cables can be cast into deck elements at the factory. They can be provided with packings to allow a quick connection at the building site.

Figure 4.18 shows an example of a bathroom floor prepared with drain, water supply, floor heating, and electricity as services built in at the factory. This saves time at the building site. It also save height, because you do not have to place a bathroom floor or bath unit on the load-bearing deck element or alternatively to place services into an in-situ cast top concrete layer.

Often, floor heating is placed as a net of pipes cast into the top of the SL-deck at the factory, where the thickness above the blocks is adjusted for the purpose. The pipes are connected to those of neighbour elements via couplers in recesses at the edge of the element.

4.2.6 Blade Connections

As mentioned in Sect. 4.1.1 and shown in Fig. 4.3, we developed blade connections already for the prototype version of the SL-deck. A blade connection is a bearing

Fig. 4.18 SL-deck with built-in services. Abeo Ltd.

detail, where one element can rest on another in the same plane without application of a beam or any other increase of the structural thickness.

A blade connection is created in a zone of strong concrete of a deck element by halving the slab thickness in a length required for the bearing.

It is necessary to provide the half thickness zone with a shear reinforcement for example a set of stirrups or a number of inclined reinforcing bars as shown in Fig. 4.22. The horizontal parts of the reinforcing bars near the bearing plane serve also to resist tension from bending of the blades.

Figure 4.19 shows a blade connection, where the end of a SL-deck element is supported on a blade inlayed in the side of another deck element. In case the last element is a SL-deck, you can omit a number of blocks, in order to make space for the blade at the side and for the reinforcement required to distribute the reaction force into the deck element.

4.3 Beamless Super-Light Decks

4.3.1 Principle of Beamless Decks

The blade connection presented in Sect. 4.2.6 allows us to establish a bearing of a slab at an end or at a side of another slab without giving rise to any increase of thickness, and without any beams. We apply this connection in the beamless deck.

We support the deck on columns, and as mentioned in Sect. 4.2.3, we need a zone of strong, reinforced concrete in a deck above a column. This is required, if it should

Fig. 4.19 Blade connection of SL-decks. *Photo* KD Hertz

transfer load from a column above to a column below the deck. As also mentioned some top reinforcement is required in two directions to resist negative moments, and perhaps some stirrups or inclined reinforcement to distribute the column reaction into the deck.

We propose using a special quadratic massive slab element with blade connections at all four sides and a width equal to the width of the SL-decks (could be 3.0 m). We place such quadratic slab element at the top of each column as shown in Fig. 4.20.

Fig. 4.20 Beamless super-light deck

Between quadratic top slabs above adjacent columns, we place a main deck element with blade connection bearings like that in Fig. 4.22 along the sides. Figure 4.20 shows this with a darker colour.

4.3.2   Moment Distribution in Beamless Decks

A number of SL-decks are supported by the blade connections at each side of the main deck element. The SL-decks are reinforced in the bottom for positive moments, and since you have zero moment at the blade connection, the main deck element is provided with a top reinforcement for the negative moments shown to the left in Fig. 4.21 in the direction across its length.

The main deck elements can be made of massive concrete, and they have bottom reinforcement in their length direction, because they only have to resist positive moments in that direction. This is due to the moment being zero in the connection to the column deck element. With such structure, you can e.g. create a beamless office building or parking garage [9]. It is an advantage that you save height, and the negative moment zones prolong the span width of decks without increasing their thickness. At the sides of the building, you may create wider top slabs to counterbalance the negative moment from the one side.

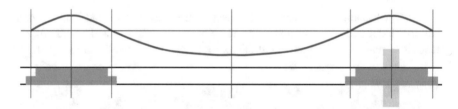

Fig. 4.21 Moment distribution in a beamless deck

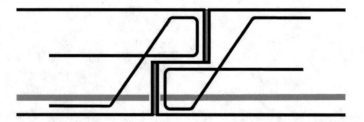

Fig. 4.22 Blade connection with shear reinforcement

4.3.3   Design for Disassembly

The blade connections are in most cases required to be poured with a mortar to ensure the building robustness. These mortar joints are only half as deep as the deck height, and this is an advantage if you should consider the disassembly and reuse of the elements. If a weak lime mortar is used, then it will be easy to remove the mortar and dismantle the building at the end of life [10].

4.4   Light Concrete Slabs

4.4.1   Massive Light Concrete Slabs

Builders have applied light aggregate concrete for slabs since antiquity.

For in-door structures, where reinforcement is not exposed to moisture, reinforced massive slabs of light aggregate concrete are quite common.

The first author has investigated fire-resistance of a number of such slab elements [11], which are representative for the production. A typical mass-produced massive deck element could be 200 mm thick, 1.2 m wide and 6 m long. It is made of a light aggregate concrete of density 1775 kg/m^3 with a compressive strength of f_{cc20} = 20 MPa, tensile strength f_{ct20} = 3.2 MPa and conductivity λ = 0.6 W/mK. It is reinforced with 8 Y 10 mm slack deformed bars of tensile strength 550 MPa with a cover thickness d = 15 mm.

The weight of the slab is about 360 kg/m^2.

The anchorage length of the deformed bars at the ends of the slab is 70 mm. We could prove by calculation as well as by test that this is sufficient according to the explanations given in Sect. 9.1.

We calculated the standard fire resistance for bending, shear and anchorage for this slab to be 69 min, and proved by test that it was 73 min for a service load of 2.4 kN/m^2 (Fig. 4.23).

Limiting parameters for designing deck structures of light aggregate concrete can be span-width, shear, anchorage, and for domestic buildings sound insulation requirements.

A density of about 1800 kg/m^3 does represent a 20% reduction of dead load compared to a similar massive slab of ordinary concrete, and this may have a little positive influence on the possible span-width. However, it has a negative influence on the sound insulation.

If you should obtain a usual minimum requirement of 55 dB, you need 440 kg/m^2 and a thickness of 250 mm. You may reduce this if you test the deck in combination with a floor structure or an in-situ cast top concrete layer.

Fig. 4.23 Fire test of 200 mm light aggregate concrete slab. *Photo* KD Hertz

4.4.2 Sandwich Slabs

If you apply a light core and strong flanges, you can obtain a light sandwich slab with a reasonable load-bearing capacity. You can typically apply such structure, where sound insulation is not required.

An example of a mass-produced sandwich deck element is 240 mm thick, 1.2 m wide and 6 m long. It is made of.

(1) a 23 mm thick top of light aggregate concrete of density 1550 kg/m^3 with compressive strength $f_{cc20} = 15.25$ MPa
(2) a 182 mm thick core of 625 kg/m^3 light aggregate concrete with and compressive strength $f_{cc20} = 2.8$ MPa and tensile strength $f_{ct20} = 0.3$ MPa
(3) a 35 mm thick bottom layer of 1500 kg/m^3 light aggregate concrete of tensile strength $f_{ct20} = 2.7$ MPa and conductivity $\lambda = 0.6$ W/mK

It is reinforced with 8 Y 8 mm slack deformed bars of tensile strength 550 MPa with a cover thickness d = 15 mm. The anchorage length is 70 mm as for the massive slab in Sect. 4.4.1.

The weight of this sandwich slab is about 205 kg/m^2.

At a fire test, the slab resisted a service load of 2.4 kN/m^2 for 79 min standard fire exposure, where it was calculated to resist it for 61 min [11].

The very light structure and low dead load allows long span, which are beneficial for many applications. However, it may cause a problem, if a sound insulation of 55 dB or more is required.

The application of more densities of concrete at the same cross-section may have a positive influence on the sound insulation. However, the sound insulation requirement may still lead to quite thick slabs even when you try to document it by testing a full floor structure.

4.5 Super-Light Concrete Beams

4.5.1 Rectangular Super-Light Beams

The first super-light structures, which we considered after inventing the principle, were ordinary rectangular beams.

When you design an ordinary concrete beam, you place the concrete and the reinforcement as usual, and then you make a structural analysis to give an idea about, how the forces approximately are distributed. Based on that, you prove that you have sufficient material in the right places of the structure to resist the forces.

As described in Sect. 3.2, the super-light principle allows you to apply direct engineering, where you at first decide an optimal distribution of forces and place a strong material such as a strong concrete or reinforcement according to that as shown in Fig. 3.4. You then fill the rest of the structure with a light concrete in order to obtain the wanted outer shape for example with a horizontal upper side of a beam, to get a fire protection of the strong materials, or perhaps a shear connection between compression- and tension zone.

As explained in Sect. 3.1.2, a curved shape of the strong material would give rise to an expensive and complicated curved mould according to the traditional way of making structures. We saw that for example when we produced the double beam in Fig. 7.10.

A simple solution to that problem could be to hang a flexible pipe upside down, cast the strong concrete in the pipe and cast it into the light concrete as shown in Fig. 3.5.

You then obtain the beam in Fig. 3.4 by turning it upside down.

Another solution could be to apply a pearl-chain structure as explained in Chap. 6 creating a curved optimized compression zone of strong concrete from a number of straight mass-produced elements. This curved structure can then be cast into a light concrete with the outer shape of the beam, which for example could be double cantilevered as shown in Fig. 7.10.

When the cross-section becomes high and slender, the beam becomes a hung-up wall, and you may apply the super-light principles and pearl-chain elements to that for example as shown in Fig. 6.6 and investigated by Lind [12]. If such high cross-sections should function as beams and not have a wall function separating rooms, you may obviously save material by providing the structure with holes, where strength is not needed as for example shown in Fig. 7.20.

High and slender beam cross-sections may also be applied as ribs in prefabricated girders for example with a double-T (TT) cross-section or cross-sections with shape of the Greek letter pi (Π).

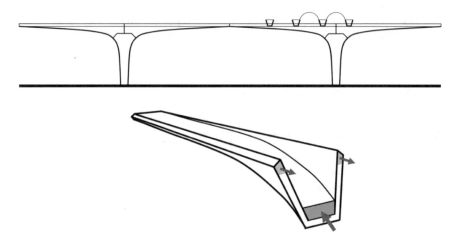

Fig. 4.24 Super-light beam system for halls

4.5.2   Super-Light Beam System for Halls

Figure 4.24 shows a principle of a system of prefabricated super-light beams for large halls. The beams are cantilevered, and two of them are connected to a double cantilevered beam for example with a length of 60 m. We tested double cantilevered beams already in 2007 [13] before the super-light principle was published and before pearl-chains were invented (Fig. 7.10). Later we applied double-cantilevered vaults for bridge engineering as shown in Figs. 7.9 and 7.17.

The super-light cantilevered beam system was at first sketched in [14]. A V-shaped cross-section for example made of a light concrete with density 1200 kg/m^3 has a curved bottom that serves as mould for casting a high-strength concrete. The curved high-strength concrete serves as a compression zone. At the top of the cross-section, we include two tension zones, which can be made for example as pretensioned zones of ordinary concrete that can resist tension as unloaded compression with the large stiffness of the concrete.

The tension zones of two half beams are bolted together, and the double cantilevered beam is placed on top of a column, where a bearing joint receives the inclined forces from the compression zones of the two half beams.

At the top of the double cantilevered beam, a number of secondary double cantilevered beams made by the same principle can be placed transversely, as shown to the right in Fig. 4.24. The figure also shows a proposal for a glazing vault between the secondary beams.

The curved bottom of the V-shaped secondary beams may serve to lead rainwater to the main beam that again may lead it to the columns, so that the roof can be drained efficiently.

References

1. Castberg NA (2013) Architectural engineering to super-light structures–design, implementation and constructability. PhD Thesis, BIG Architects and Department of Civil Engineering, Technical University of Denmark, 213p
2. Tassello A (2011) Load-bearing capacity of super-light slabs. MSc Project, Department of Civil Engineering, Technical University of Denmark, 72p
3. Hertz KD, Castberg Christensen AJ (2014) Super-light concrete decks for building floor slabs. Struct Concr J FIB 15(4):522–529. Ernst & Sohn
4. Halldórsson EE (2012) Load bearing capacity of super-light decks. MSc Thesis, Faculty of Civil and Environmental Engineering, University of Iceland, 60p
5. Rocca N (2010) Risposta al fuoco di elementi strutturali superleggeri. MSc Thesis, Università degli Studi di Roma La Sapienza, 209p
6. Hertz KD (2019) Design of fire-resistant concrete structures. ICE Publishing, Thomas Telford Ltd. ISBN: 9780727764447. London, 254p
7. Christensen JE (2013) Acoustic design of super-light structures. PhD Thesis, Gronmij Ltd and Department of Civil Engineering, Technical University of Denmark. 140p
8. ISO (International Organization for Standardization) (1975) ISO 834 Fire-resistance tests. Elements of building construction. ISO Geneva, Switzerland, 25p
9. Persson K, Raaschou L (2017) Super-light multi-story car park. MSc Project, Department of Civil Engineering, Technical University of Denmark, 230p
10. Halding PS, Hertz KD (2020) Design for disassembly of super-light structures. RILEM spring convention 2020
11. Hertz KD (2003) Documentation for calculations of standard fire resistance of slabs and walls of concrete with expanded clay aggregate. Report R-048. Department of Civil Engineering, Technical University of Denmark. ISBN 87–7877–108–0. Lyngby, 43p
12. Lind FL (2018) Super-light wall elements. MSc Project, Department of Civil Engineering, Technical University of Denmark, 82p
13. Larsen F (2007) Light concrete structures. MSc Project, Department of Civil Engineering, Technical University of Denmark, 117p
14. Hertz KD (2009) Super-light concrete with pearl-chains. Mag Concr Res 61(8):655–663. Thomas Telford Ltd.

Chapter 5
Columns and Walls

Abstract The chapter provides an in depth method with examples on how to calculate the capacity of columns or walls using layers of different types of concrete. The force in a column can be guided via a strong concrete, while a lighter concrete stabilizes the strong against buckling. Columns with Entasis are used as well.

5.1 Light Concrete Columns and Walls with Uniform Cross-Section

5.1.1 Centrally Loaded Light Concrete Columns and Walls

Light concrete columns and walls can be made from light concrete with or without steel reinforcement, but may also be composite and consist of parts of strong and light concrete.

The composite cross-section may consist of several areas of strong and light concrete. Figure 5.1 shows a simple example with a strong concrete section in the centre. By placing a light concrete around the strong part, the designer may apply the light concrete for stabilizing and protecting the strong concrete.

You will find this especially relevant if you apply a high-strength concrete susceptible to explosive spalling in fire [1], where the light concrete keeps the temperature below the required maximum of 350 °C. Furthermore, the light concrete avoids buckling the small section that will be a result of application of a high strength concrete.

In this Sects. 5.1 and 5.2, we deal with columns, where the cross-section is uniform along the length of the column. This has been the usual way of designing columns since the re-introduction of reinforced concrete in modern time, because it is made with simple moulds, which were easy to construct and therefore economically optimal.

However, it is not the most optimal solution statically to apply columns with uniform cross-sections. Varying cross-sections may give better load-bearing capacities, and with a new light concrete technology, where flexural textile moulds make

Fig. 5.1 Centrally loaded
light composite concrete
column

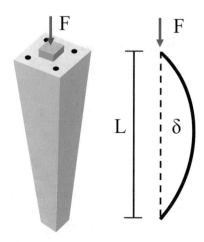

it affordable, such structures will often be economical and sustainable. We therefore
deal with these structures in Sect. 5.3.

For a plain column of one material, Euler (Fig. 5.2) presumed an elastic sinusoidal
deflection when he made his well-known formula for the load-bearing capacity of a
centrally loaded column.

$$y = \delta \sin\left(\frac{\pi}{L}x\right) \Rightarrow \frac{dy}{dx} = \frac{\pi}{L}\delta \cos\left(\frac{\pi}{L}x\right) \Rightarrow$$

$$\frac{d^2y}{dx^2} = \kappa = -\frac{\pi^2}{L^2}\delta \sin\left(\frac{\pi}{L}x\right),$$

Moment equilibrium (inner moment = outer moment) of the cross-section at the
middle, where $x = L/2$, gives for an applied force F, and the deflection at the middle
δ

Fig. 5.2 William John Macquorn Rankine, Wilhelm Ritter, and Leonhard Euler. (Drawing by KD
Hertz)

$$\delta F = -\kappa \, E_0 I \Leftrightarrow \delta F = \frac{\pi^2}{L^2} \delta E_0 I \Leftrightarrow$$

$$F_E = \frac{\pi^2 E_0 I}{L^2}$$

In this derivation y is the elastic deflection in the length x.

κ is the curvature, L is the column length, and I is the moment of inertia of the cross-section.

We have here presumed that the material is linear elastic.

From the derivation, you see that we not only obtain equilibrium between outer and inner moment at the middle, but at any level x of the column.

$$y F_E = -\kappa(x) E_0 I \Leftrightarrow \delta \sin\left(\frac{\pi}{L}x\right)\left(\frac{\pi^2 E_0 I}{L^2}\right) = \frac{\pi^2}{L^2}\delta \sin\left(\frac{\pi}{L}x\right)E_0 I$$

Euler's expression represents equilibrium of the moment from the outer force times the deflection and the inner elastic moment given as a product of flexural stiffness (EI) and curvature.

As such, it only represents a requirement of stability and not a requirement of sufficient compressive strength, which a designer must also consider, especially when the slenderness of the column decreases.

As seen from the formula, the Euler force varies towards infinity, when the column length L varies towards none. This is obviously wrong. Rankine (Fig. 5.2) therefore suggested applying a variation limited to the compressive strength of the material. Ritter (Fig. 5.2) then explained this by describing equilibrium of a deflected column with a certain load for a material that has a curved lined stress–strain curve as seen in Fig. 5.3 [2].

Fig. 5.3 Stress–strain curve according to Ritter

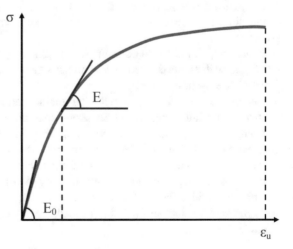

We will obtain equilibrium for that if we apply the related reduced elastic modulus equal to the inclination of the tangent to the stress–strain curve at that particular load level.

We find this E-modulus from Ritter's assumption

$$E = E_0\left(1 - \frac{\sigma}{f_u}\right) = E_0\left(1 - \frac{F}{F_u}\right) \text{ or } \frac{d\sigma}{d\varepsilon} = E_0\left(1 - \frac{\sigma}{f_u}\right)$$

In this expression, F_u represents the ultimate load in compression for the cross-section and F the actual load on the column, which we want to find.

Solving the differential equation above, Ritter's stress–strain curve becomes

$$\sigma(\varepsilon) = f_u\left(1 - e^{-\left(\frac{E_0}{f_u}\right)\varepsilon}\right)$$

We call the force at equilibrium in the deflected column for the Rankine force F_R.

$$F_R = \frac{\pi^2 I E_0}{L^2}\left(1 - \frac{F}{F_u}\right) \Leftrightarrow F_R = F_E\left(1 - \frac{F_R}{F_u}\right)$$

From that, we get the simple expression

$$\frac{1}{F_R} = \frac{1}{F_u} + \frac{1}{F_E}$$

F_R is the ultimate load for the centrally loaded column.

This is a convenient expression for the so-called Rankine formula applied in most textbooks and codes on concrete structures, although it usually appears as expressions, which are hard to recognize.

Still, we have only assumed an ultimate strength, a Ritter stress–strain curve, and a sinusoidal deflection as a basis for our structural calculations. We have derived everything else.

In general, all parts of a column (strong concrete, light concrete, and reinforcement) follow the same deflection, and all parts including modern reinforcement have a curved stress–strain curve.

Hence, it is possible for you to find a Rankine load-bearing capacity for a column consisting of reinforcing bars alone, strong concrete alone or light concrete alone.

Doing so, you presume that something keeps the mutual distance between the different parts when they deflect. This means that you may add the contributions of the cross-sectional areas of each material each weighted by its elastic modulus.

Ritter also added steel and concrete areas by multiplication of the contribution of the steel area with a factor 10 that was an approximation for E_{s0}/E_{c0} and f_{su}/f_{cu}. We apply the actual values of E_0 and f_u instead, when we add contributions of areas of different materials.

From that, you may get an extended Rankine formula for a centrally loaded reinforced composite concrete column combining the three expressions.

This approach is similar to the expressions developed in [1] for calculating the load-bearing capacity of fire exposed reinforced concrete columns. In these, the impact of fire has changed the cross-sections so that they contain a variety of concrete qualities and a variety of reinforcement qualities dependent on how much the fire has heated each zone of a cross-section.

Here, s means steel, c strong concrete and lc light concrete.

$$\frac{1}{F_R} = \frac{1}{F_{cu} + F_{lcu} + F_{su}} + \frac{1}{F_{cE} + F_{lcE} + F_{sE}}$$

This expression shows the load-bearing capacity F_R of a centrally loaded reinforced light composite concrete column.

5.1.2 Eccentrically Loaded Uncracked Columns and Walls

In the previous clause, we derived the Rankine formula for the load-bearing capacity of a centrally loaded reinforced composite concrete column.

However, we apply eccentric load to most columns as illustrated in Fig. 5.4.

We will therefore check the stresses in a deflected, eccentrically loaded column to see whether it is stable for a given load F, and eccentricity e.

Here, we apply the following expressions for the total axial- (EA) and flexural stiffness (EI) of a composite cross-section at a certain load level related to a common axis of bending

Fig. 5.4 Eccentrically loaded column

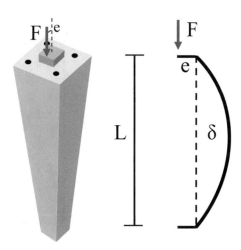

$$EA = E_c A_c + E_{lc} A_{lc} + E_s A_s \text{ and } EI = E_c I_c + E_{lc} I_{lc} + E_s I_s$$

For an unloaded cross-section the same properties become

$$EA_0 = E_{c0} A_c + E_{lc0} A_{lc} + E_{s0} A_s \text{ and } EI_0 = E_{c0} I_c + E_{lc0} I_{lc} + E_{s0} I_s$$

The ultimate compressive strength of the total cross-section is also needed in the formulas.

For an ultimate strain of concrete of more than 0.35% and the steel yielding above $0.2\% + f_{su}/E_{su} =$ approximately 0.4%, you may assume the strain to be the same and

$$F_u = A_c f_{cu} + A_{lc} f_{lcu} + A_s f_{su}$$

We apply the tangent elastic module of the curved working curve suggested by [2] for each of the concrete qualities in the cross-section. We find the tangent elastic modulus as inclination of the tangent to the curved working curve for the strain of the entire cross-section subjected to the axial central load F.

We can then find a stress-distribution for the eccentricity e using the tangent elastic modulus of each area (Fig. 5.5) and thereby we assume that all parts of the cross-section deform according to their position and that plane cross-sections remain plane.

We make the reasonable presumption that the entire cross-section will fail when the strongest part fail. We therefore consider each of the different concrete qualities to follow a Ritter stress–strain curve with the same ultimate strain. This means that we consider the ratio of the reduction of the elastic moduli to be the same for all types of concrete, when we apply a load F, on the column. In case the concretes have

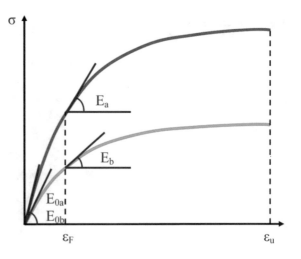

Fig. 5.5 Stress–strain curves for different concretes in the same cross-section

different ultimate strains, we apply the smallest, which will also be the one of the strongest concrete.

We can therefore find the reduced elastic modulus for each of the different materials as

$$E = E_0\left(1 - \frac{\sigma}{f_u}\right) = E_0\left(1 - \frac{F}{F_u}\right)$$

This means that we can apply this reduction to the entire axial and flexural stiffness.

$$EA = E_c A_c + E_{lc} A_{lc} + EA_s \Leftrightarrow$$

$$EA = (E_{c0} A_c + E_{lc0} A_{lc} + E_{s0} A_s)\left(1 - \frac{F}{F_u}\right) \Leftrightarrow$$

$$EA = EA_0\left(1 - \frac{F}{F_u}\right)$$

And similarly

$$EI = EI_0\left(1 - \frac{F}{F_u}\right)$$

The Euler strength of the total composite cross-section then becomes

$$F_E = \frac{\pi^2 EI_0}{L^2}$$

To assess the curvature and deflection δ for the eccentric loaded column, we apply the same sinusoidal deflection curve as Euler because we, just like Euler, claim that the difference of the result between using that and the actual curve will be modest. We then apply a condition of equilibrium similar to the one Euler used for a central loaded column but now with a lever arm, $e + \delta$, to the force, F, in a deflected condition at the midpoint (Fig. 5.4). This means that we require the external moment of the force with lever arm $e + \delta$ to be equal to the internal moment of the cross-section subjected to the curvature for the deflection δ.

$$(e + \delta)F = \frac{\pi^2}{L^2}\delta\, EI \Leftrightarrow (e + \delta)F = \frac{\pi^2 EI_0}{L^2}\left(1 - \frac{F}{F_u}\right)\delta \Leftrightarrow$$

$$e\frac{F}{F_E} = \left(1 - \frac{F}{F_u} - \frac{F}{F_E}\right)\delta \Leftrightarrow e\frac{F}{F_E} = \left(1 - \frac{F}{F_R}\right)\delta \Leftrightarrow$$

$$\delta = e\frac{\frac{F}{F_E}}{1 - \frac{F}{F_R}} \Leftrightarrow (e + \delta) \Leftrightarrow e\frac{1 - \frac{F}{F_R} + \frac{F}{F_E}}{1 - \frac{F}{F_R}}\delta(e + \delta) \Leftrightarrow e\frac{1 - \frac{F}{F_u}}{1 - \frac{F}{F_R}}$$

This expression calculates the lever arm e + δ in a deflected condition, of a column loaded by a force F with an eccentricity e. The eccentricity e is increased by a sinusoidal deflection of magnitude δ in the midpoint in equilibrium with the force. We can then check the load-bearing capacity of a composite column for a given force F by checking that the total cross-section can withstand the moment (e + δ) F and the normal force F. If that is the case, the column can carry the normal force F with an initial eccentricity e with respect to instability.

$$(e + \delta) = e \frac{1 - \dfrac{F}{F_u}}{1 - \dfrac{F}{F_R}}$$

If we know the normal force F and eccentricity e, we have to design the cross-section of the column to resist F and moment (e + δ) F. In case we want to calculate the load-bearing capacity of a given column with a given eccentricity, we have to guess the force F and check the cross-section for that, and within a couple of iterations find the critical F.

In both cases, we can apply different methods for calculating the composite cross-section just as we can for beams with a normal force.

(1)   We can find an elastic stress-distribution and compare the stresses at the edges with the strength of the materials in compression and tension.
      This is usually on the safe side compared to a plastic approach.
(2)   We can find a plastic stress-distribution presuming that no tensile stresses are applied.
(3)   We can find a plastic stress-distribution with application of tensile stresses.

This is applicable for beam-columns, where the bending moment is large.

In the first case, we calculate stresses at the edges of each part of the cross-section such as strong concrete, light concrete and reinforcement, for the column subjected to a certain load F and moment (e + δ) F and compare the stresses with the strength. As an example, for the strong concrete we compare the compressive stress at the distance y_{cc} from the centroid of the composite cross-section with the compressive strength f_{cc}

$$\sigma_{c1} \le f_{cc} \Leftrightarrow \frac{F E_c}{EA} + \frac{F(e + \delta)E_c}{EI} y_{cc} \le f_{cc} \Leftrightarrow$$

$$\frac{F E_{c0}\left(1 - \frac{F}{F_u}\right)}{EA_0\left(1 - \frac{F}{F_u}\right)} + \frac{F(e + \delta)E_{c0}\left(1 - \frac{F}{F_u}\right)}{EI_0\left(1 - \frac{F}{F_u}\right)} y_{cc} \le f_{cc} \Leftrightarrow$$

$$\boxed{\frac{FE_{c0}}{EA_0} + \frac{F(e + \delta)E_{c0}}{EI_0} y_{cc} \le f_{cc}}$$

Similarly, we compare the tensile stress at the distance y_{ct} with the tensile strength f_{ct}

$$\sigma_{c2} \geq -f_{ct} \Leftrightarrow \frac{F E_c}{EA} - \frac{F(e + \delta)E_c}{EI} y_{ct} \geq -f_{ct} \Leftrightarrow$$

$$\frac{F E_{c0}\left(1 - \frac{F}{F_u}\right)}{E_0 A\left(1 - \frac{F}{F_u}\right)} - \frac{F(e + \delta)E_{c0}\left(1 - \frac{F}{F_u}\right)}{E_0 I\left(1 - \frac{F}{F_u}\right)} y_{ct} \geq -f_{ct} \Leftrightarrow$$

$$\boxed{\frac{F E_{c0}}{EA_0} - \frac{F(e + \delta)E_{c0}}{EI_0} y_{ct} \geq -f_{ct}}$$

We then do the same check for light concrete and reinforcement.

5.1.3 Columns and Walls with Plastic Compression

In this case, we consider a column or wall with height L and a massive cross-section of for example a light concrete. We presume that the cross-section should be able to resist a force with eccentricity e in a deflected condition by a plastic distribution of ultimate compressive stresses symmetrically placed around the eccentric force.

We denote the height of the concrete cross-section h in the direction of the eccentricity e. The width is always c no matter if the height is larger or not.

In order to estimate the deflection we calculate the compressive strength F_u and the flexural stiffness EI_0 using the initial elastic modulus E_0 as shown in Sect. 5.1.1.

From that, we find the Euler force

$$F_E = \frac{\pi^2 EI_0}{L^2}$$

The result will be a Rankine force F_{Rc} from the formula

$$\frac{1}{F_R} = \frac{1}{F_u} + \frac{1}{F_E}$$

The total lever arm at the middle of the column is then the sum of the eccentricity e and the deflection δ. In Sect. 5.1.2 we found an expression for $(e + \delta)$ as

$$(e + \delta) = e\frac{1 - \frac{F}{F_u}}{1 - \frac{F}{F_R}}$$

If we want to find the critical load F for a column or wall as shown in Fig. 5.6, we can presume a value of F in order to find the deflection δ. The calculation is an iteration.

Fig. 5.6 Plastic compression
on a massive column or wall

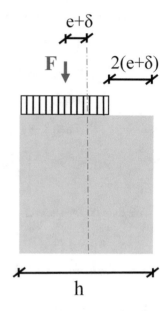

However, when we apply the expression for the lever arm $(e + \delta)$ for calculation of
the compressive strength symmetrically around the eccentric force F we can obtain
an explicit expression of the critical load. This means that we can avoid iteration

$$\frac{F}{F_u} = \frac{h - 2(e + \delta)}{h} \Leftrightarrow \frac{F}{F_u} = 1 - 2\left(\frac{e}{h}\right)\frac{1 - \frac{F}{F_u}}{1 - \frac{F}{F_R}} \Leftrightarrow$$

$$1 - \frac{F}{F_R} = 2\frac{e}{h} \Leftrightarrow$$

$$\boxed{F = F_R\left(1 - 2\frac{e}{h}\right)}$$

5.1.4 Beam-Columns and Cracked Walls

We call the structure a beam-column, if we apply a moment load so large that the
tension zone of the concrete cracks and we mainly rely on resisting tension in the
reinforcement.

This means that the column cannot carry its load as an uncracked structure
according to Sects. 5.1.2 and 5.1.3.

For uncracked columns, we considered the stiffness of the entire cross-section
calculating the deflection and we found the ultimate moment capacity of the part of
the cross-section with strong concrete and reinforcement. This is in accordance with

the calculations we make for the ultimate load-bearing capacity of composite beams and slabs and for uncracked composite columns.

For cracked beam-columns, we also find the ultimate moment resistance of the strong concrete alone and compare this with the moment in the column at the ultimate deflection, where the strains in the cross-section correspond to the ultimate moment.

The column load F has a depth d_F from the compressed edge of the strong concrete. If the column force has an eccentricity e from a centroid in a depth y_{cc}, this is

$$d_F = y_{cc} - e$$

However, all you need to know is the depth d_F of the column load before deflection.

The depth d_F is positive if the column load F acts within the strong concrete before the column deflects and negative, if the moment is so large that it acts beyond the compressed edge.

We call the width of the strong concrete across the direction of the depth d_F for c_c.

c_c does not need to be the smallest dimension of the strong concrete, but often it is.

The strong concrete may also consist of more sections and then c_c is the width of the section with the compressed edge, where the compressive stresses will be in a cracked column.

For a beam-column with tension in the tensile reinforcement, we can assess the ultimate deflection δ_u of a reinforced cross-section from the curvature calculated for the ultimate resistance of the cross-section at the critical column load F.

We apply the ultimate strain for normal concrete $\varepsilon_{cu} = 0.0035$ at the compressed edge of the strong concrete, and the yield strain ε_{sy} and the yield force of a tension reinforcement F_{s2} in the depth d_{s2} from the compressed edge of the strong concrete (Fig. 5.7).

Fig. 5.7 Beam-column cross-section with eccentric load F before deflection

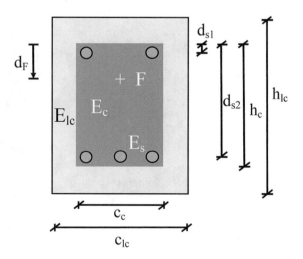

$$\varepsilon_{sy} = 0.002 + \frac{f_s}{E_{s0}}$$

Still presuming a sinusoidal deflection, we find the ultimate value δ_u of this deflection as

$$\delta_u = \frac{\varepsilon_{cu} + \varepsilon_{sy}}{\pi^2 d_{s2}} L^2$$

The deflection δ_u reduces the depth of the column load from the compressed edge, which means that the deflection increases the moment of the column load. In some cases, the column force F may act beyond the compressed edge and its depth d_F is negative. In such cases, the deflection still reduces the lever arm and increases its negative value.

For the ultimate deflection, the external moment load of the critical column load F calculated about the compressed edge of the strong part of the cross-section becomes

$$M = -F(d_F - \delta_u)$$

Reinforcement will yield in compression before the concrete around it fails. Therefore, we can also obtain a yield force F_{s1} in the compression reinforcement at depth d_{s1} near the compressed edge (Fig. 5.8).

Fig. 5.8 Cracked composite
column with ultimate
deflection δ_u

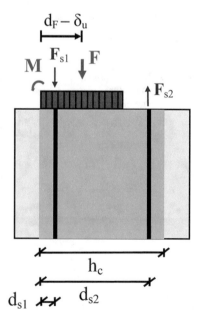

The total force in the reinforcement is

$$F_s = F_{s2} - F_{s1} \quad \text{(Positive in tension)}$$

The moment contribution about the compressed edge of F_{s2} in tension and F_{s1} in compression is then

$$\underline{M_s = d_{s2}F_{s2} - d_{s1}F_{s1}}$$

You then find the moment contribution about the compressed edge of the force $F_c = F + F_s$ in the strong concrete section of width c_c and compressive strength f_{cc}

$$M_c = -(F + F_s)\frac{(F + F_s)}{2c_c f_{cc}}$$

Inserting these expressions in the condition that the external moment of the load should be equal to the internal moment in the deflected cross-section, you find a second-order equation in the critical column load F

$$M = M_c + M_s \Leftrightarrow -M = -M_c - M_s \Leftrightarrow$$
$$(d_F - \delta_u)F2c_c f_{cc} = \left(F^2 + 2FF_s + F_s^2\right) - M_s 2c_c f_{cc} \Leftrightarrow$$
$$F^2 + (2F_s - (d_F - \delta_u)2c_c f_{cc})F + F_s^2 - 2c_c f_{cc}M_s = 0$$

In order to solve this, we define a parameter b and find the critical load F as

$$\boxed{\begin{array}{l} b = 2F_s - (d_F - \delta_u)2c_c f_{cc} \\[2mm] F = \dfrac{-b + \sqrt{b^2 - 4F_s^2 + 8c_c f_{cc}M_s}}{2} \end{array}}$$

This is an expression of the critical load F of an eccentrically loaded reinforced concrete column with tensile stresses in the reinforcement in a deflected condition – a beam-column.

You can then compare the result of F with the applied load F_{load} and the eccentricity measured as the depth d_F of the load from the compressed edge of the strong concrete before deflection.

5.1.5 Design of Columns and Walls—Summary

When we design an eccentric loaded column or wall, we can chose how we want the column to react to the maximum column force in the deflected condition.

If we want that the column or wall to be safe form cracking, we apply method (1).

If a massive column or wall should carry the load in compression, we apply method (2).

If we allow cracks and ultimate stresses in strong concrete and reinforcement, we apply method (3).

Method (1). Uncracked according to Sect. 5.1.2. Here we apply a column force F that we either know or guess with an eccentricity e.

We find the ultimate compressive load and the flexural stiffness

$$F_u = A_c f_{cu} + A_{lc} f_{lcu} + A_s f_{su} \quad \text{and} \quad EI_0 = E_{c0} I_c + E_{lc0} I_{lc} + E_{s0} I_s$$

Using the ultimate compressive strength f_u and the initial modulus of elasticity E_0 for each material, we find the Euler force F_E and from that, the Rankine force F_R

$$F_E = \frac{\pi^2 EI_0}{L^2} \quad \text{and} \quad \frac{1}{F_R} = \frac{1}{F_u} + \frac{1}{F_E}$$

The Euler force F_E is the critical load of the column or wall, if the eccentricity is $e = 0$ mm and the materials are linear elastic. The Rankine force F_R is the critical load if we use the tangent elastic moduli of the stress–strain curves at the load level considered.

From that, we find the lever arm equal to the initial eccentricity plus the deflection $e + \delta$:

$$(e + \delta) = e \frac{1 - \dfrac{F}{F_u}}{1 - \dfrac{F}{F_R}}$$

We then check that the stresses do not exceed the strength anywhere in the deflected composite cross-section as for example the maximum tensile stress of the strong concrete at the edge in tension at the distance y_{ct} from the centroid of the composite cross-section

$$\frac{F E_{c0}}{EA_0} - \frac{F(e + \delta) E_{c0}}{EI_0} y_{ct} \geq -f_{ct}$$

We apply the initial elastic modulus for each material E_0 to find the initial axial and flexural stiffness EA_0 and EI_0. We do that, because we have proved that the tangential elastic modulus and the axial and flexural stiffness of the loaded cross-section would give the same result.

This is because the stiffness of the different materials relate equally to each other in a loaded and in an unloaded cross-section.

Method (2). In case we consider a massive column or wall, we can avoid guessing a critical load F, if we consider a plastic uncracked cross-section according to Sect. 5.1.4.

The capacity is then:

$$F = F_R \left(1 - 2\frac{e}{h} \right)$$

Method (3). A cracked cross-section according to Sect. 5.1.4.

Here, we find the ultimate plastic moment of the strong concrete and the rein-forcement, as we would do for a beam. We apply a deflection δ_u corresponding to the ultimate plastic moment:

$$\delta_u = \frac{\varepsilon_{cu} + \varepsilon_{sy}}{\pi^2 d_{s2}} L^2$$

Thereby we find the ultimate critical load of the column in the cracked condition, where we consider it as a beam with normal load F in a depth before deflection d_F from the compressed edge of the strong concrete and with a width c_c. We calculate d_F as positive for F in the cross-section and negative for larger moments, where F is placed beyond the compressed edge.

$$b = 2F_s - (d_F - \delta_u) 2 c_c f_{cc}$$

$$F = \frac{-b + \sqrt{b^2 - 4F_s^2 + 8c_c f_{cc} M_s}}{2}$$

Here, we do not apply the strength and stiffness of the light concrete (as we also do not for plastic moments for beams. The result is therefore on the safe side for the entire column.

5.2 Examples Columns and Walls with Uniform Cross-Section

5.2.1 Example of a Tested Column

As an example, we consider a $L = 2.0$ m long composite concrete column (Fig. 5.9) with a central core of strong concrete h_c is 50 mm and c_c is 90 mm. The ultimate compressive strength f_{cc} is 45 MPa, the tensile strength f_{ct} is 3.8 MPa, and the initial elastic modulus E_{c0} is 36 GPa. The column is hinged in both ends, and it has been full-scale tested by [3].

Around the central strong core is cast a light aggregate concrete of density 950 kg/m^3 with cross-sectional properties: $h_{lc} = 180$ mm, $c_{lc} = 190$ mm, ultimate strength $f_{lcc} = 5.0$ MPa, tensile strength $f_{lct} = 0.5$ MPa and initial elastic modulus $E_{lc0} = 5.0$ GPa (Fig. 5.10).

Fig. 5.9 Composite column test

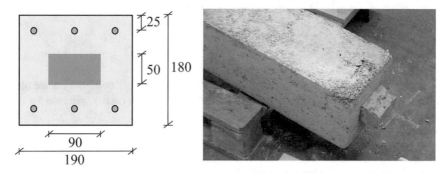

Fig. 5.10 Cross-section and cracked composite column

Six Y10 reinforcing bars with $D_s = 10$ mm, yield strength $f_s = 550$ MPa, and elastic modulus $E_{s0} = 200$ GPa are placed in the light aggregate concrete with cover thickness $c_s = 25$ mm.

Their area A_s, distance from centre y_s, and moment of inertia I_s are

$$A_s = 6\frac{\pi}{4}D_s^2, \quad y_s = \frac{h_{lc}}{2} - c_s, \quad I_s = A_s y_s^2$$

A load is applied with an eccentricity $e = 10$ mm. We calculate the load-bearing capacity and compare it to the capacity found in the laboratory test.

The area and moment of inertia of both the strong concrete and light aggregate concrete and the steel is

$$A_c = h_c c_c, \quad I_c = \frac{1}{12} c_c h_c^3$$

$$A_{lc} = h_{lc} c_{lc} - A_c, \quad I_{lc} = \frac{1}{12} c_{lc} h_{lc}^3 - I_c$$

The ultimate strength of the strong- and light concrete and the steel is

$$F_{cc} = A_c f_{cc} = 203 \text{ kN}, \quad F_{lcc} = A_{lc} f_{lcc} = 149 \text{ kN}, \quad F_s = A_s f_s = 259 \text{ kN}$$

The total ultimate strength is

$$F_u = F_{cc} + F_{lcc} + F_s = 610 \text{ kN}$$

The axial stiffness of the column is

$$EA_0 = E_{c0} A_c + E_{lc0} A_{lc} + E_{s0} A_s = 405 \text{ MN}$$

The flexural stiffness is

$$EI_0 = E_{c0} I_c + E_{lc0} I_{lc} + E_{s0} I_s = 889 \text{ kN m}^2$$

The Euler strength of the strong- and light concrete and steel is

$$F_{cE} = \frac{\pi^2 EI_{c0}}{L^2} = 83 \text{ kN}, \quad F_{lcE} = \frac{\pi^2 EI_{lc0}}{L^2} = 1128 \text{ kN}, \quad F_{sE} = \frac{\pi^2 EI_{s0}}{L^2} = 983 \text{ kN}$$

The total Euler strength is $F_E = F_{cE} + F_{lcE} + F_s = 2193 \text{ kN}$.
The Rankine force of the column becomes

$$F_R = \frac{1}{\frac{1}{F_u} + \frac{1}{F_E}} = 477 \text{ kN}$$

If we guess a column force $F = 266$ kN, we can calculate the corresponding deflection, where the external moment from the axial load with eccentricity and deflection is in equilibrium with the internal moment with an elastic distribution of stresses at the composite cross-section.

$$(e + \delta) = e \frac{1 - \frac{F}{F_u}}{1 + \frac{1}{F_R}} = 12.74 \text{ mm}$$

With this balancing moment, we can find the maximum and minimum stresses at the edges of the strong and the light concrete and in the reinforcing bars.

$$\sigma_{c1} = \frac{FE_{c0}}{EA_0} + \frac{F(e+\delta)E_{c0}}{EI_0}\frac{h_c}{2} = 27.1 \text{ MPa} \quad \le f_{cc} = 45 \text{ MPa} \quad \checkmark$$

$$\sigma_{c2} = \frac{FE_{c0}}{EA_0} - \frac{F(e+\delta)E_{c0}}{EI_0}\frac{h_c}{2} = 20.2 \text{ MPa} \quad \ge -f_{ct} = -3.8 \text{ MPa} \quad \checkmark$$

$$\sigma_{lc1} = \frac{FE_{lc0}}{EA_0} + \frac{F(e+\delta)E_{lc0}}{EI_0}\frac{h_{lc}}{2} = 5.00 \text{ MPa} \quad \le f_{lcc} = 5 \text{ MPa} \quad \checkmark$$

$$\sigma_{lc2} = \frac{FE_{lc0}}{EA_0} - \frac{F(e+\delta)E_{lc0}}{EI_0}\frac{h_{lc}}{2} = 1.57 \text{ MPa} \quad \ge -f_{lct} = -0.5 \text{MPa} \quad \checkmark$$

$$\sigma_{s1} = \frac{FE_{s0}}{EA_0} + \frac{F(e+\delta)E_{s0}}{EI_0}y_s = 181.0 \text{ MPa} \quad \le f_s = 550 \text{MPa} \quad \checkmark$$

$$\sigma_{s2} = \frac{FE_{s0}}{EA_0} - \frac{F(e+\delta)E_{s0}}{EI_0}y_s = 81.9 \text{ MPa} \quad \ge -f_s = -550 \text{MPa} \quad \checkmark$$

This means that the column can carry at least 266 kN with the eccentricity 10 mm, if the force acts on the total composite cross-section and we consider an elastic stress distribution according to Sect. 5.1.2.

If we instead only consider the load-bearing capacity according to Sect. 5.1.4 for a cracked cross-section with ultimate moment in the reinforcement and strong concrete. Then we get $\delta_u = 37.2$ mm, $F_s = 0$ kN, $M_s = 16.9$ kNm

$$\varepsilon_{sy} = 0.2\% + \frac{f_s}{E_{s0}} = 0.00475; \varepsilon_{cu} = 0.00350; \quad d_{s2} = 0.09 \text{ m};$$

$$\delta_u = \frac{\varepsilon_{cu} + \varepsilon_{sy}}{\pi^2 d_{s2}}L^2 = 0.0372 \text{ m}; \quad d_F = \frac{h_c}{2} - e = 0.015 \text{ m}$$

$$M_s = d_{s2} F_{s2} - d_{s1} F_{s1} = (0.09 \text{ m} - (-0.04 \text{ m}))F_{s2} = 16.9 \text{ kNm}$$

and

$$b = 2F_s - (d_F - \delta_u)2c f_{cc} = 179 \text{ kN}$$

$$F = \frac{-b + \sqrt{b^2 - 4F_s^2 + 8c f_{cc}M_s}}{2} = 290 \text{ kN}$$

We can therefore expect the load-bearing capacity of the column to be 266 kN before it cracks and 290 kN in an ultimate cracked condition.

However, the tested column was constructed so that the force acted on a part of the central strong concrete section sticking out of the column as seen in Fig. 5.10.

We therefore have to find the minimum load-bearing capacity as a minimum of the critical load of the column and the strength of this part subjected to the eccentric load without deflection.

Fig. 5.11 Failed strong column

The capacity in pure eccentric compression of the projected part will be

$$F_{cc}\left(1 - \frac{2e}{h_c}\right) = 121.5 \text{ kN}$$

This capacity is less than the critical load of the column and therefore determines the capacity of the whole structure at the test.

We made two tests with such column [3], and saw failures at 122 kN and 128 kN. The failures occurred at the projecting strong concrete in both tests, and no damage was observed on the rest of the columns (Fig. 5.10).

Then, we tested a column consisting of the central, unreinforced, strong concrete part alone without the enclosing, stabilizing light aggregate concrete (Fig. 5.11). We still applied a load with eccentricity e = 10 mm. The Rankine force of this column is:

$$F_{cR} = \frac{1}{\dfrac{1}{F_{cc}} + \dfrac{1}{F_{cE}}} = 59 \text{ kN}$$

The deflection, where the internal moment is in equilibrium with the external is

$$(e + \delta) = e\frac{1 - \dfrac{F}{F_{cc}}}{1 - \dfrac{1}{F_{cR}}} = 14.53 \text{ mm}$$

For a safe-side calculation based on maximum and minimum stresses of an elastic distribution we can guess a capacity of F = 24 kN, and then we get

$$\sigma_{c1} = \frac{F}{A_c} + \frac{F(e + \delta)}{I_c} \frac{h_c}{2} = 14.02 \text{ MPa} \quad \le f_{cc} = 45 \text{ MPa}$$

$$\sigma_{c2} = \frac{F}{A_c} - \frac{F(e + \delta)}{I_c} \frac{h_c}{2} = -3.8 \text{ MPa} \quad \ge -f_{ct} = -3.8 \text{ MPa}$$

If we instead find the ultimate load-bearing capacity of the column based on a plastic stress-distribution in the deflected condition as shown in Sect. 5.1.3, we get:

$$F_{cR}\left(1 - \frac{2e}{h_c}\right) = 35.4 \text{ kN}$$

We should therefore expect that the tested column would have a failure load between 24 kN and 35.4 kN. In the test, the column failed at 29.5 kN (Fig. 5.11).

As seen from the example, the composite column, where a light concrete stabilizes the strong, had a load-bearing capacity that was four times larger than the same column consisting of only the strong concrete. Furthermore, if we had not applied the load on a weak projecting part of the strong concrete, we could expect it to have been nine times stronger.

We will now consider the composite column loaded with varying eccentricity and without failure at the projecting strong part. With an increasing eccentricity from 0 to 60 mm, we find the corresponding load-bearing capacity considering the uncracked- or the cracked condition.

Figure 5.12 shows the result. When the eccentricity increases to 25 mm, the critical condition for the uncracked column changes from crushing of the compression zone of the light concrete to cracking of the tension zone of the light concrete. You can see that as a change in inclination of the curve at this point.

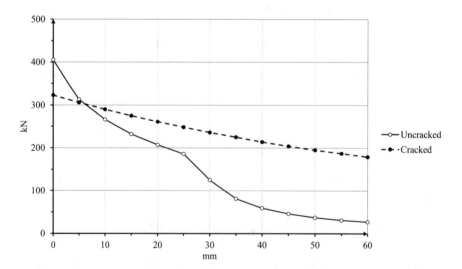

Fig. 5.12 Theoretical capacity of test column with variable eccentricity

For eccentricities less than 5 mm, the column will fail as soon as it reaches the limit for the uncracked condition. For larger eccentricities, the column can deflect further and carry a larger force than the limit for the uncracked condition.

5.2.2 Super-Light Wall Design

Lind [4] investigated a number of designs of super-light wall elements comprising sandwich walls as described in Sect. 5.2.3 and walls, where strong concrete constituted frames with holes filled out by light concrete.

The main problem for sustainable light wall elements is usually a requirement of sound insulation for separating walls. The Danish Building Regulations requires for example 55 dB, and this means that a massive wall should have a mass of 440 kg/m^2. Super-light sandwich walls may fulfill the requirement with a mass of only 350 kg/m^2 because the concrete materials of different density oscillates with different eigen-frequencies, transforming some of the sound energy to heat, as we can see from the sound tests on super-light deck elements described in Sect. 4.1.4.

For this reason, sandwich walls or sandwich like walls may represent more promising design solutions for domestic buildings than walls designed as frames of strong concrete with light fillings. However, they can be interesting, where sound requirements do not represent a design limit.

5.2.3 Example Sandwich Wall

We now consider a part c = 1 m of a L = 3.0 m high sandwich wall (Fig. 5.13) consisting of two layers of strong concrete (each t_c = 30 mm thick) separated by a layer of light concrete of thickness h_{lc} = 90 mm.

The wall thickness is then h_c = 2·30 + 90 = 150 mm.

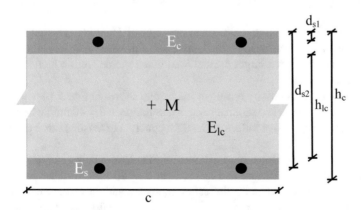

Fig. 5.13 Sandwich wall cross-section

The strong concrete has an ultimate compressive strength $f_{cc} = 55$ MPa, a tensile strength $f_{ct} = 4.2$ MPa, and an initial elastic modulus $E_{c0} = 38$ GPa.

The light aggregate concrete is of density 700 kg/m^3 and has an ultimate strength $f_{lcc} = 3.0$ MPa, a tensile strength $f_{ct} = 0.3$ MPa, and an initial elastic modulus $E_{cl0} = 3.0$ GPa.

An Y8 reinforcement ($D_s = 8$ mm) with the strength $f_s = 550$ MPa is placed per 150 mm in each "flange" of strong concrete.

For these data input we get: $F_{cc} = 3300$ kN, $F_{lcc} = 270$ kN and $F_s = 369$ kN.

In addition, the Euler force becomes

$$F_{cE} = 9189 \text{ kN}, F_{lcE} = 200 \text{ kN} \quad \text{and} \quad F_{sE} = 529 \text{ kN. In total } F_E = 9918 \text{ kN.}$$

It is observed that the two flanges of strong concrete are the main load-carrying parts of the cross-section.

We find the ultimate compressive force and the Rankine force

$$F_u = 3939 \text{ kN}, F_R = 2819 \text{ kN.}$$

For an eccentricity of e = 20 mm we guess an uncracked capacity of F = 1890 kN.

From that we find get $(e + \delta) = 31.6$ mm and check the maximum stresses

$$\sigma_{c1} = 45.6 \text{ MPa} \quad \leq f_{cc} = 55 \text{ MPa}$$
$$\sigma_{c2} = 7.96 \text{ MPa} \quad \geq f_{ct} = -4.2 \text{ MPa}$$
$$\sigma_{lc1} = 3.00 \text{ MPa} \quad \leq f_{lcc} = 3.0 \text{ MPa}$$
$$\sigma_{lc2} = 1.22 \text{ MPa} \quad \geq f_{lct} = -0.3 \text{ MPa}$$
$$\sigma_{s1} = 220.0 \text{ MPa} \quad \leq f_s = 550 \text{ MPa}$$
$$\sigma_{s2} = 61.7 \text{ MPa} \quad \geq f_s = -550 \text{ MPa}$$

The guessed force is the limit for compressive failure of the light concrete.

Furthermore, we find the cracked deflection $\delta_u = 55.7$ mm, $M_s = 22.1$ kNm and b = 79.9 kN.

From that, we determine the cracked capacity of 1520 kN. This means that the sandwich wall with eccentricity of 20 mm will fail when it is loaded to crushing, because it does not get a larger critical load when the deflection increases and the cross-section cracks.

If we vary the eccentricity e, we get the results shown in Fig. 5.14.

The curve for uncracked failure changes inclination at e = 40 mm, where the limit changes from compressive failure of the light concrete to tensile failure of the strong concrete.

For this wall the reinforcement will increase the load-bearing capacity for eccentricities less than e = 10 mm and larger than e = 80 mm. In these intervals, the critical load increases, when the deflection increases and the cross-section cracks.

In between these limits the wall will break, when the first crushing occurs.

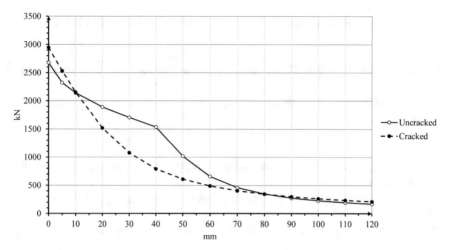

Fig. 5.14 Theoretical capacity of sandwich wall with variable eccentricity

5.2.4 *Example Wall and Column with Circular Light Cores*

If the central light concrete filling of a sandwich wall is replaced with a series of circular light concrete cores as shown in Fig. 5.15, you obtain the benefit of a sandwich structure and in addition an improved axial stiffness. The profile may be beneficial for columns with one or more circular cavities.

We now consider a part $c = 1$ m of a $L = 3.0$ m high wall with mainly the same data as the sandwich wall in Example 5.2.3 $t_c = 30$ mm (thickness flanges) and circular light concrete parts with diametre $D_{lc} = 90$ mm. The total cross-sectional height is still $h_c = 2{\cdot}30 + 90 = 150$ mm.

As for the sandwich wall we apply a strong concrete with $f_{cc} = 55$ MPa, $f_{ct} = 4.2$ MPa and $E_{c0} = 38$ GPa. The light concrete of density 700 kg/m^3 has $f_{lcc} = 3.0$ MPa, $f_{lct} = 0.3$ MPa and $E_{cl0} = 3.0$ GPa.

Y8 reinforcement $D_s = 8$ mm has the strength $f_s = 550$ MPa and is placed per 120 mm (each segment is a 90 mm circle plus a 30 mm web equal to t_c).

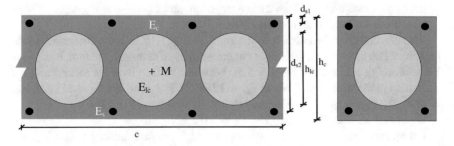

Fig. 5.15 Wall and column with circular light cores

For these data we get: $F_{cc} = 5334$ kN, $F_{lcc} = 159$ kN and $F_s = 461$ kN. The Euler force is

$$F_{cE} = 10,602 \text{ kN}, F_{lcE} = 88 \text{ kN} \quad \text{and} \quad F_{sE} = 662 \text{ kN. In total } F_E = 11,352 \text{ kN.}$$

From that we determine the ultimate compression force and the Rankine force

$$F_u = 5954 \text{ kN}, F_R = 3906 \text{ kN.}$$

The strong concrete is the main load-carrying part of the cross-section as it also was for the sandwich wall.

For an eccentricity of e = 20 mm we guess an uncracked capacity of F = 2550 kN.

From that we find get $(e + \delta) = 32.9$ mm and check the maximum stresses.

$$\sigma_{c1} = 47.3 \text{ MPa} \quad \leq f_{cc} = 55 \text{ MPa}$$
$$\sigma_{c2} = 1.02 \text{ MPa} \quad \geq f_{ct} = -4.2 \text{ MPa}$$
$$\sigma_{lc1} = 3.00 \text{ MPa} \quad \leq f_{lcc} = 3.0 \text{ MPa}$$
$$\sigma_{lc2} = 0.81 \text{ MPa} \quad \geq f_{lct} = -0.3 \text{ MPa}$$
$$\sigma_{s1} = 224.5 \text{ MPa} \quad \leq f_s = 550 \text{ MPa}$$
$$\sigma_{s2} = 29.7 \text{ MPa} \quad \geq f_s = -550 \text{ MPa}$$

The guessed force is the limit for compression failure of the light concrete.

Furthermore, we find the cracked deflection $\delta_u = 55.7$ mm, $M_s = 27.7$ kNm and b = 79.9 kN.

From that, we find the cracked capacity of 1704 kN. This means that the wall with eccentricity of 20 mm will fail in case it is loaded to crushing, because an increase of deflection will not increase the critical load.

If we vary the eccentricity e, we get the results shown in Fig. 5.16.

When the eccentricity increases to more than e = 30 mm, tension failure of the strong concrete happens before compression failure of the light concrete.

This wall has a larger load-bearing capacity at the ultimate deflection for all eccentricities than the similar sandwich wall in Example 5.2.3. It has a larger cross-section of strong concrete, and therefore it has a larger uncracked load-bearing capacity than the sandwich wall for small eccentricities.

We now consider the column to the right in Fig. 5.15. It is quadratic with side lengths of 200 mm and a central light concrete cavity of diameter 140 mm. It has four Y8 reinforcing bars and length L = 3 m. We consider it to have the same materials as applied in the wall. This leads to $F_{cc} = 1353$ kN, $F_{lcc} = 46$ kN, $F_s = 111$ kN, $F_{cE} = 4770$ kN, $F_{lcE} = 62$ kN and $F_{sE} = 319$ kN. In total $F_E = 5151$ kN.

From that we find: $F_u = 1510$ kN, $F_R = 1168$ kN.

For an eccentricity of e = 20 mm we guess an uncracked capacity of F = 722 kN and from that we get $(e + \delta) = 27.3$ mm.

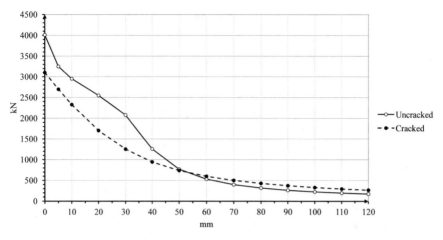

Fig. 5.16 Theoretical capacity of wall with circular light cores and variable eccentricity

From the eccentricity e = 50 mm the failure mode in the uncracked condition changes from compression in the light concrete to tension in the strong concrete.

At all eccentricities, the cracked condition has a larger capacity, which means that the column can be loaded beyond the limit, where the first cracking is observed (Fig. 5.17).

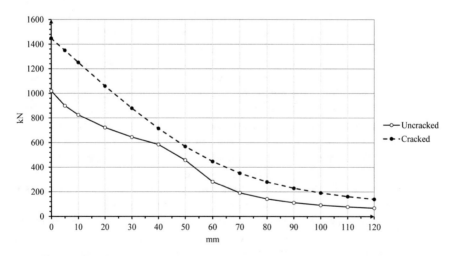

Fig. 5.17 Capacity of column with circular light core and variable eccentricity

5.3 Light Concrete Columns and Walls with Entasis

5.3.1 Centrally Loaded Columns and Walls with Entasis

Entasis is according to Vitruvius [5] the Greek expression for a light convex curved surface of a column, so that it is thicker at the middle than at the ends. This gives an optical effect counteracting the impression that a straight column may appear more slender at the middle, and many authors refer to this as a reason why ancient architects applied it.

An engineering reason may be that a column with Entasis has the largest stiffness and cross-sectional resistance at the middle, where the deflection and the curvature in a deflected condition is the largest for a simply supported column. In other words, application of Entasis counteracts deflection by placing material, where the column requires it the most and thereby serve a purpose of reducing the total material consume.

The Greek word Entasis means "tension" and is formed of "en- " and "teinein", which means to stress. This is probably because a column with Entasis gives an optical impression of tension or swelling. However, for a light concrete structure this meaning may become physical. If we cast a concrete column with Entasis in a mould, this "swelling" gives a tensile stress in the mould material and the convex curved shape becomes natural for counteracting the mould pressure in tension.

Furthermore, application of a light concrete reduces the mould pressure and the tensile stress for example to a quarter of that of an ordinary concrete. This means that we can apply quite different mould materials such as light textiles and obtain more elegant, economical, and sustainable constructions.

From a point of view of engineering optimization, we would like to utilize all cross-sections of the column better than we do for an ordinary straight column, where the cross-section at the middle is subjected to the largest curvature.

As explained in Sect. 5.1.1, you derive the Euler force based on a presumption that the deflection of a simply supported column with a uniform cross-section follows a sinus curve. This implies that the curvature of the column is zero at the ends and maximum at the midpoint. It is a reasonable presumption since the moment in a centrally loaded column is proportional to the deflection.

However, when you design your column with an entasis, you obtain a reduction of the curvature at the midpoint compared to a column with uniform cross-section because you have increased the flexural stiffness. Since most columns are calculated for an eccentric load, it would be more reasonable to apply a presumed deflection with a uniform curvature κ along the height of the columns with entasis so that all parts of the column contribute equally much to resist curvature and deflection.

This presumption gives us a precondition for the deflection $\delta - y(x)$, where δ is the deflection at the middle of the column, where we consider $x = 0$ m. The ends of the column are then at $x = L/2$ and $x = - L/2$, where $y(L/2) = \delta$. (Fig. 5.18)

Fig. 5.18 Circular column
with Entasis

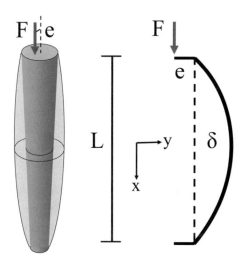

$$\frac{d^2y}{dx^2} = \kappa \Leftrightarrow \frac{dy}{dx} = \kappa x + a \Leftrightarrow y(x) = \frac{1}{2}\kappa x^2 + a\,x + b$$

Symmetry gives $a = 0$ and the precondition that $y(0) = 0$ gives $b = 0$ and we get:

$$y(x) = \frac{1}{2}\kappa x^2$$

$$y\left(\frac{L}{2}\right) = \frac{1}{2}\kappa\left(\frac{L}{2}\right)^2 \Leftrightarrow \kappa = 8\frac{\delta}{L^2}$$

At first, we consider the materials linear elastic with elastic flexural stiffness $E_0 I(x)$. External equals internal moment gives

$$F(\delta - y(x)) = E_0 I(x) \Leftrightarrow F\left(\delta - 4\frac{x^2}{L^2}\delta\right) = 8\frac{\delta}{L^2}E_0 I(x)$$

For $x = 0$ you get the load-bearing capacity of a centrally loaded column or wall with Entasis, linear elastic material and constant curvature

$$F_0 = \frac{8E_0 I(0)}{L^2}$$

This value is 81% of the similar Euler force, and as explained above, it is more reasonable to apply for eccentric loaded columns with Entasis.

As for the Euler force in Sect. 5.1.1 for columns with uniform cross-section, it is now possible to replace the linear elastic materials with materials with curved working curves according to the Ritter assumption. We consider the flexural stiffness at the middle of the column (where $x = 0$ m) and reduce the elastic modulus in that for the load of the Rankine force F_{R0} of the column or wall with entasis. This gives us

$$F_{R0} = \frac{8IE_0}{L^2}\left(1 - \frac{F}{F_u}\right) \Leftrightarrow F_{R0} = F_0\left(1 - \frac{F_{R0}}{F_u}\right)$$

From that, we get the simple expression

$$\frac{1}{F_{R0}} = \frac{1}{F_u} + \frac{1}{F_0}$$

As for columns and walls with uniform cross-sections, we can consider the cross-sections as composite structures with

$$F_u = A_c f_{cu} + A_{lc} f_{lcu} + A_s f_{su} \quad \text{and} \quad EI_0 = E_{c0}I_c + E_{lc0}I_{lc} + E_{s0}I_s$$

Similar to columns and walls with uniform cross-section, we can derive an expression for the total deflection of a column or wall with an Entasis and a constant curvature.

$$(e + \delta)F = \frac{8}{L^2}\delta\, EI(0) \Leftrightarrow (e + \delta)F = \frac{8E_0I(0)}{L^2}\left(1 - \frac{F}{F_u}\right)\delta \Leftrightarrow$$

$$e\frac{F}{F_0} = \left(1 - \frac{F}{F_u} - \frac{F}{F_0}\right)\delta \Leftrightarrow e\frac{F}{F_0} = \left(1 - \frac{F}{F_{R0}}\right)\delta \Leftrightarrow$$

$$\delta = e\frac{\dfrac{F}{F_0}}{1 - \dfrac{F}{F_{R0}}} \Leftrightarrow (e + \delta) = e\frac{1 - \dfrac{F}{F_{R0}} + \dfrac{F}{F_0}}{1 - \dfrac{F}{F_{R0}}} \Leftrightarrow$$

$$(e + \delta) = e\frac{1 - \dfrac{F}{F_u}}{1 - \dfrac{F}{F_{R0}}}$$

The eccentricity at the ends of the column is e. It will be $(e + \delta)$ in a deflected condition at the midpoint, when the column is subjected to the axial load F.

$$\boxed{(e + \delta) = e\frac{1 - \dfrac{F}{F_u}}{1 - \dfrac{F}{F_{R0}}}}$$

You can then check that the maximum stresses in the cross-sections at the midpoint and at the top and bottom are not exceeding the ultimate stresses. For the strong concrete

$$\boxed{\frac{F E_{c0}}{EA_0} + \frac{F(e+\delta)E_{c0}}{EI_0}\frac{h_c}{2} \le f_{cu}}$$

$$\boxed{\frac{F E_{c0}}{EA_0} - \frac{F(e+\delta)E_{c0}}{EI_0}\frac{h_c}{2} \ge -f_{ctu}}$$

h_c is the thickness of the strong concrete. You do the similar check for the light concrete and for possible reinforcement.

If we know the normal force F and the eccentricity e, we have to design the cross-section of the column to be able to resist the moment Fe at the end and $F(e+\delta)$ at the midpoint.

5.3.2 Example of a Column with Entasis

In this example, we consider a column with a uniform circular core of strong concrete of diametre $D_c = 160$ mm with compressive strength $f_{cc} = 45$ MPa, tensile strength $f_{ct} = 3.8$ MPa and initial elastic modulus $E_{c0} = 36$ GPa.

The circular core is surrounded by a circular light aggregate concrete section with no thickness at top and bottom and thickness 60 mm at the midpoint (Fig. 5.18). This means that the light combined section has a outer diametre of $D_m = 280$ mm and an inner diametre of 160 mm at the midpoint.

The light aggregate concrete has a density of 950 kg/m^3, ultimate strength $f_{lcc} = 5.0$ MPa, tensile strength $f_{lct} = 0.5$ MPa and initial elastic modulus $E_{lc0} = 5.0$ GPa.

The column has a height $L = 3$ m and is loaded by the force $F = 300$ kN with an eccentricity $e = 20$ mm.

$$A_c = \frac{\pi}{4}D_c^2, \quad I_c = \frac{\pi}{32}D_c^4, \quad A_{lc} = \frac{\pi}{4}D_m^2 - A_c, \quad I_{lc} = \frac{\pi}{32}D_m^4 - I_c$$

The ultimate strength of the strong concrete and the total ultimate strength are

$$F_{cc} = A_c f_{cc} = 905 \text{kN}, \quad F_u = F_{cc} + A_{lc}f_{lcc} = 1112 \text{ kN}$$

The axial stiffness of the column is

$$EA_0 = E_{c0}A_c + E_{lc0}A_{lc} = 931 \text{ MN}$$

The flexural stiffness is

$$EI_0 = E_{c0}I_c + E_{lc0}I_{lc} = 2506 \text{ kN m}^2$$

The centrally loaded strength of the column is

$$F_0 = \frac{8EI_{c0}}{L^2} = 2227 \text{ kN}$$

The Rankine force of the column becomes

$$F_{R0} = \frac{1}{\dfrac{1}{F_u} + \dfrac{1}{F_0}} = 742 \text{ kN}$$

The total deflection at the midpoint becomes

$$(e + \delta) = e \frac{1 - \dfrac{F}{F_u}}{1 - \dfrac{F}{F_{R0}}} = 24.5 \text{ mm}$$

Checking the stresses at the edges of the light aggregate concrete at the midpoint you get

$$\sigma_{lc1} = \frac{FE_{lc0}}{EA_0} + \frac{F(e + \delta)E_{lc0}}{EI_0} \frac{D_m}{2} = 3.67 \text{ MPa} \quad \leq f_{lcc} = 5.0 \text{ MPa} \ \checkmark$$

$$\sigma_{lc2} = \frac{FE_{lc0}}{EA_0} - \frac{F(e + \delta)E_{lc0}}{EI_0} \frac{D_m}{2} = -0.44 \text{ MPa} \quad \geq -f_{lct} = -0.5 \text{ MPa} \ \checkmark$$

In addition, at the edges of the strong concrete at the midpoint you get

$$\sigma_{c1} = 20.1 \text{ MPa} \leq f_{cc} = 45 \text{ MPa} \checkmark$$
$$\sigma_{c2} = 3.14 \text{ MPa} \geq f_{ct} = -3.8 \text{ MPa} \ \checkmark$$

At the top the strong concrete should carry the load alone, and here we find

$$\sigma_{c1top} = 29.8 \text{ MPa} \leq f_{cc} = 45 \text{ MPa} \ \checkmark$$
$$\sigma_{c2top} = 0.00 \text{ MPa} \geq f_{ct} = -3.8 \text{ MPa} \ \checkmark$$

This means that the column can carry the load 300 kN with the eccentricity 20 mm.
We now remove the light aggregate concrete and thereby the entasis and consider the strong concrete part of the column alone:

$$F_{cE} = \frac{\pi^2 E_{c0} I_c}{L^2} = 1270 \text{ kN}, \quad F_{cR} = \frac{1}{\dfrac{1}{F_{cc}} + \dfrac{1}{F_{cE}}} = 528 \text{ kN}$$

$$(e + \delta) = e \frac{1 - \dfrac{F}{F_{cc}}}{1 - \dfrac{F}{F_{cR}}} = 30.9 \text{ mm}$$

And from that

$$\sigma_{c1m} = 38.0 \text{ MPa} \quad \leq f_{cc} = 45 \text{ MPa} \quad \checkmark$$
$$\sigma_{c2m} = -8.15 \text{ MPa} \geq f_{ct} = -3.8 \text{ MPa} \quad \text{Failure !}$$

As seen, the strong concrete alone cannot carry the load 300 kN.
It can only carry a load of 233 kN. This is the effect of the Entasis.

References

1. Hertz KD (2019) Design of fire-resistant concrete structures. ICE Publishing, Thomas Telford Ltd. ISBN: 9780727764447. London, 254p
2. Ritter W (1899) Die Bauweise Hennebique. (In German). Schweizerische Bauzeitung vol 33, Heft 7, pp 59–61
3. Larsen F (2007) Light concrete structures. MSc Project, Department of Civil Engineering, Technical University of Denmark. 117p
4. Lind FL (2018) Super-light wall elements. MSc Project, Department of Civil Engineering, Technical University of Denmark. 82p
5. Vitruvius M (-35) De Architecture. Several publications: Om Arkitektur (In Danish). Translation edited by J Isager. University of Southern Denmark 2016, 505p
6. CEN (2006) EN 1992-1-2 Eurocode 2, Design of concrete structures, Part 1-2. General rules—structural fire design. Brussels. 97p
7. CEN (2008) EN 1992-1-1 Eurocode 2, Design of Concrete Structures, Part 1-1. General rules and rules for buildings. Brussels. 225p
8. The ten books on Architecture (In English). Translated by MH Morgan. Dover 1960. Harvard University Press 1914, 325p

Chapter 6
Pearl-Chain Structures

Abstract Pearl-chain reinforcement is introduced, where pre-fabricated concrete segments are post-tensioned together in any desired shape. Pearl-chain arches have been tested and applied in full-scale, and it is shown how different types of concrete segments can be used in many variations in roofs, vaults, bridges, cantilevered slabs etc.

6.1 Pearl-Chains

6.1.1 Pearl-Chain Principle

Arches and vaults are optimal structures capable of carrying loads in pure compression.

As described in Chap. 1 on history, builders have applied these structures for thousands of years, and benefitted from large spans, low material consume, elegant appearance, and no tension that requires expensive materials.

However, it takes labour to construct them. Usually you need a curved scaffolding and a curved mould for erection of an arch or vault, and labour has become expensive in the modern society. Consequently, many engineers have not applied these structures for half a century. In addition, materials applied for scaffolding and moulds represent extra consume of resources and extra impact on the climate.

Nevertheless, arches and vaults are still geometrically optimal structures with respect to resisting distributed loads. In many cases they represent the best and most economical solutions with the smallest material consume and climate impact if they could only be built without application of scaffolding or curved moulds.

This is the reason why, the first author has invented the pearl-chain principle [1].

By means of this, you can create any curved arch or vault structure from a series of elements without application of scaffolding.

The single elements can often be equal or consist of very few types and they can be mass-produced. In many cases, the elements can even be straight or plane and therefore easy to produce and easy to transport from factory to building site as demonstrated by the test pearl-chain arch shown in Fig. 6.1.

K. D. Hertz and P. Halding., *Sustainable Light Concrete Structures*, Springer Tracts in Civil Engineering, https://doi.org/10.1007/978-3-030-80500-5_6

Fig. 6.1 Pearl-chain test arch. *Photo* KD Hertz

The idea came from a little toy animal as the one shown in Fig. 6.2.

When you put pearls on a string and pull the string, you introduce tension forces in the string and compression forces in the pearls. The compression forces have the same centroid as the tension forces in the string.

Because you have placed the string in tubular holes in the pearls you can place the compression forces, where you want them, for example at the centroid of the cross-section of the pearls.

This means that if a chain of pearls forms a curved shape due to inclined ends of the pearls, the tension in the string will preserve the curved shape.

The principle is used in the toy animal, where the animal rises and the pearl-chains become one rigid unit, when the string is tightened.

You may also apply a curved path of the holes between the ends of each pearl. By doing so, you can avoid that the string should pass edges at the assemblies. In addition, you can obtain a curved path of the prestressing force, which may be optimal for resisting distributed lateral forces on each element.

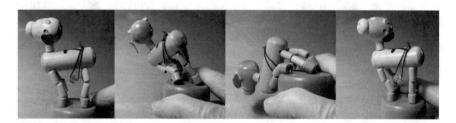

Fig. 6.2 Pearl-chain toy. *Photo* KD Hertz

6.1.2 Prestressing

Eugène Freyssinet (Fig. 6.3) invented prestressed concrete structures [2]. His idea was to tension straight or curved reinforcement cables, wires, or rods placed in ducts in a structure, and anchor them to the concrete introducing similar compression forces in the concrete sections.

Again, the tension and the compression forces are equal. Their centroids and directions are the same—also if the cables are curved. This is because the cable and the concrete mutually interact with lateral stresses changing the direction of the tension as well as the compression force.

In mass-produced concrete elements such as the SL-deck, prestressing is applied as *pre-tensioned* strands. This means that you tension a number of prestressing strands in a prestressing bed (a mould), before pouring the concrete. Each strand usually consist of a single wire, and the prestressing bed is often 100 or 150 m long. You then cut the strands between the elements before they are de-moulded. This process transfers the prestressing forces to the concrete of each element.

In general, this production method requires some large foundations to resist the prestressing force during casting, and therefore you mainly produce pre-tensioned concrete at factories and mainly as straight pre-tensioned strands.

For in-situ cast structures *post-tensioned* strands are applied. Here, you cast ducts into the concrete. After casting, you pull strands consisting of a number of wires through the ducts and tension and anchor them to the ends of the concrete. This gives good opportunities for designing prestressing strands with curved geometries that can counteract actual moment and shear distributions [3].

Figure 6.4 shows an example of this. The first author has designed a 40 m long half-cylindrical shell supported on three walls to carry the load of a roof to both sides. It is provided with six post-tensioned strands in each side with a shape of a double suspension bridge.

Fig. 6.3 Eugène Freyssinet. Drawing KD Hertz

Fig. 6.4 Post-tensioned half-cylindrical shell. *Photo* KD Hertz

The structure can resist tension as unloaded compression along these strands. By means of that, moments and shear from a maximum service load without safety factors (service limit state) will not give rise to cracking. Additional slack reinforcement at the bottom of the shell resists the difference in force between the ultimate load (including safety factors) and the service load.

This kind of design, where you apply prestressing to avoid cracking and deflection for service load and allow cracking for ultimate load is known as partial prestressing [4–6].

In the half-cylindrical shell, the engineer used prestressing not only to counteract deflections, but mainly to avoid cracking for shear forces in the slender sides of the shell.

100 years of experience with prestressing technology means that builders and structural engineers are familiar with the technology required to pull strands through a concrete structure and introduce a desired compression.

The benefits of prestressing are well known:

(1)   To resist tension as unloaded compression.
(2)   To avoid cracking.
(3)   To keep the full stiffness of the cross-section when loaded in tension or by a moment.
(4)   To counteract deflection by application of curved or eccentric prestressing.

When using prestressing techniques for establishing a pearl-chain structure of several concrete elements, you obtain the same benefits.

In addition, you can obtain the following main advantages for curved structures:

(1)   To avoid expensive curved moulds.
(2)   To avoid expensive curved scaffoldings.

(3)   To apply inexpensive prefabricated elements.
(4)   To apply equal straight elements, which are inexpensive to produce and transport.

6.1.3 Pearl-Chain Elements

Figure 6.5 shows how a curved structure can be established using straight elements with a small inclination of the ends. Each element can contain one or more ducts for the prestressing lines. The ducts should usually be perpendicular to each inclined end surface in order to ensure a continuous shape, which is necessary when pulling the strands.

You would often place the centroid of the ducts centrally in the cross-section of the element. This gives a uniformly distributed compressive prestress at the cross-section by preventing bending moments.

If the pearl-chain is an arch, the prestress will be supplemented with the compressive stresses from the main load of the arch. On top of these, you can then add or subtract bending stresses from secondary skew loads on the arch. Tension is then resisted as unloaded compression. This means that neither the cross-sections of the elements nor the joints between them may crack [7].

In case you for some reason decide to make a pearl-chain that does not function as a perfect arch, which is shaped to support the main loads in compression, you may compensate for this by placing the prestressing ducts with an eccentricity in the pearl-chain elements. This is similar to what you do in other post-tensioned structures as explained in Sect. 6.1.2.

Even for a pearl-chain, which is supposed to act as a perfect compression arch, you may want to counteract temporary moments, due to lifting.

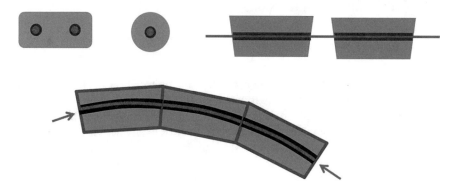

Fig. 6.5 Curved pearl-chain made from straight elements

Fig. 6.6 Tension pearl-chain

The pearl-chain principle is convenient to create curved structures of concrete easily at a low cost, and with a low environmental impact. It is even more beneficial to apply pearl-chain arches with super-light cross-sections of strong concrete embedded, stabilized, and protected by light concrete.

By combining super-light structures [8] and the pearl-chain principle [1], you have the best possible opportunities for placing strong concrete exactly, where you want the forces to be, and by that design your structure becomes optimal and more economical.

Prestressing of light concrete has seldom been possible, because the light concrete may creep when exposed to prestress. By means of super-light pearl-chains, you can embed prestressed load-bearing parts of strong concrete and at the same time create a light prestressed concrete structure.

Figure 6.6 is a special example of this. Here a light concrete wall is hung up in a tension pearl-chain, where tension is resisted as unloaded compression. The figure shows the pearl chain supported on two columns.

6.2   Pearl-Chain Applications

6.2.1   Pearl-Chain Arches

Figure 6.7 shows two flat super-light deck elements prepared for the first full-scale test of a pearl-chain arch. The end surfaces are cast with an inclination of 3.5° and a retarder is applied to the end moulds so that the aggregates are visible. This creates a rough surface that can transfer shear from one element to another via mortar joints. In addition, two recesses are made to assist transferring shear between the elements [9–11].

As a standard, we have designed SL-deck elements with spacing between the light concrete blocks at the centre. This spacing is designed to allow room for a post-tensioning cable for pearl-chain structures (see Sect. 4.1.2), and we therefore positioned the cable duct here. The duct consists of a standard flexible metal tube, where the prestressing wires are to be pulled. The tube is placed perpendicular to

Fig. 6.7 SL-deck pearl-chain elements. *Photo* Abeo Ltd.

the inclined end surfaces of the element [12]. This position was simply obtained by placing a single piece of wood under the middle of the tube before casting the element. The stiffness of the tube then provided a perfect curve between the ends.

Figure 6.8 shows the first full-scale pearl-chain arch lifted in the middle element for negative moments and Fig. 6.9 the same subjected to positive moments.

Fig. 6.8 First full-scale test of a pearl-chain arch. *Photo* KD Hertz

Fig. 6.9 Load test of first pearl-chain arch. *Photo* KD Hertz

6.2.2 Integrated Pearl-Chains

Designing two-dimensional constructions like plates, slabs, vaults, and shells, you may appreciate an opportunity to concentrate forces in arches, ribs, or beams in a pearl-chain as an integrated part of the structure.

You can apply such a pearl-chain in a larger construction as a skeleton resisting the main forces. It can also serve to support moulds for the light concrete in a super-light structure. Often, such integrated pearl-chains are preferable to apply in edges of a plate, slab, vault, or shell to make the edges robust and to ensure the strong parts the largest possible lever arm.

If you divide the larger construction into structural elements, you can include the segments of the pearl-chain as parts of larger elements as shown in Fig. 6.10.

In many cases, integrated pearl-chains serve as arches, ribs, or straight stringer components transferring axial loads as compression or tension. The pearl-chain

Fig. 6.10 Integrated pearl-chain

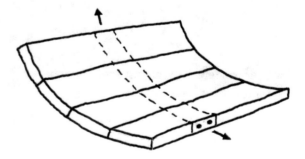

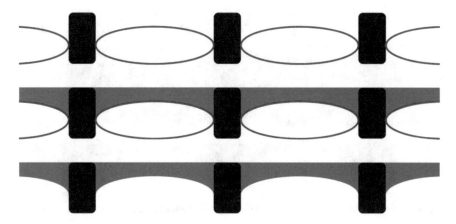

Fig. 6.11 Building with ribs and balloon moulds

prestressing allows you to resist tension as unloaded compression and thereby avoid cracks and benefit from the full axial stiffness of the cross-section of the pearl-chain.

In addition, you have the same benefits, when you design a pearl-chain for secondary moment loads, which may appear for example due to skew live loading.

This means that you can apply integrated pearl-chains as ribs of strong concrete embedded in a light concrete or you can make the rib structure thicker and visible from the inside with geometry like the one of classical rib construction known from many cathedrals.

To do so, you may for example establish a system of ribs as pearl-chains of strong concrete and between them cast a shell of light concrete in situ. Since the light concrete has an about four times smaller density than the strong (and thereby a four times smaller mould pressure), the moulds for casting it can be made light, simple, and inexpensive by means of for instance textiles, which even can be supported by balloons as shown in Fig. 6.11.

A method for such construction could be:

(1) Build the rib structure.
(2) Blow up balloons between the ribs, where the balloon pressure ensures that the moulds can fit geometrical irregularities and make the mould tight.
(3) Cast the light concrete between the ribs.
(4) Remove the balloons, and you can benefit from nice vaulted bottom surfaces of the light concrete between the ribs.
(5) You may cast a thin layer of strong concrete above the ribs, in case the light concrete is not sufficient to transfer the shear forces.

Alternatively, you can create a ribbed structure of purely factory made elements.

A fully prefabricated design could be to create a frame of strong concrete partly or fully filled out with a light concrete plate or vault as shown in Fig. 6.12.

Each rib then consists of frame sides from two neighbour elements with a joint in full height of the rib that is convenient to cast out.

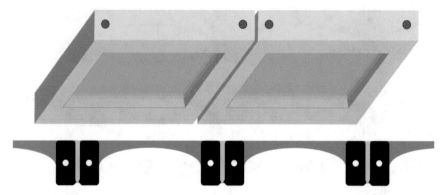

Fig. 6.12 Two dimensional elements for a ribbed pearl-chain structure

You can include ducts for prestressing in these edge frames, and this allows you to post-tension a series of such cassette elements as a pearl-chain to be lifted in place as a whole and to function as a whole in the final structure.

In addition, you may provide the transverse parts of the frames with ducts allowing you to post-tension the elements across the direction of the primary pearl-chain.

This is relevant for cupolas designed as radial pearl-chains of trapezoidal elements.

Here, you usually require a transverse post-tensioning of the pearl-chains in order to provide a ring compression force that can counteract the lateral forces from the cupola.

See more about cupola design in Sect. 8.1.

6.2.3   Cross-Stringer Structures

A cross-stringer structure consists of straight integrated stringers in two directions connected by shear plates. This structure has proved to be effective as a model for designing shear walls since the first author developed it in 1976 [13, 14].

For massive walls, you can postulate a distribution of forces in vertical and horizontal stringers. You may choose the stiffness of a stringer to that of the wall next to it as the stiffness of the total cross-section if the wall is in compression or the stiffness of the reinforcement if it is cracked. A slack reinforcement stringer can be everything from a single bar to a bundle of for example nine corrugated bars. The system of equilibrium equations for shear and stringer forces gives a ribbon matrix, and as soon as your computer program has solved it, you can easily check the dimensions.

Figure 6.13 shows an example of a folded wall. The stringers crossing the fold have zero stiffness and the same length on both sides of the fold. This ensures that only shear is transferred across the fold.

Figure 6.14 shows a cross-stringer model of a wall that acts as a high continuous beam and the long horizontal openings meant a special challenge for the design.

Fig. 6.13 Stringer model of
a folded wall

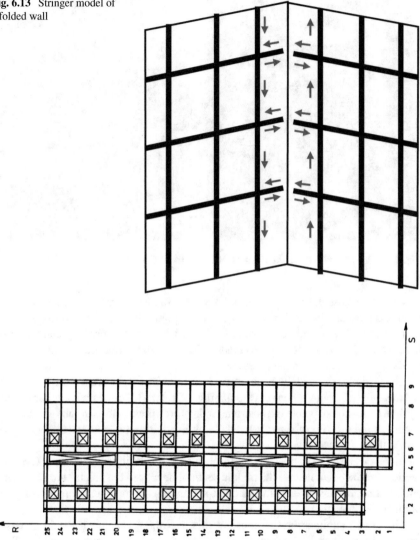

Fig. 6.14   Stringer model of a complicated wall

6.2.4   Integrated Edge-Beams

You may often need a one-dimensional structure at the edge of a two-dimensional
wall, slab or shell—capable of concentrating forces or supporting the edge in bending
either in the plane of the structure or out of it.

Figure 6.15 shows such 30 m long edge-beam designed by the first author.

Fig. 6.15   Prestressed beam at the Royal Danish Theatre in Copenhagen. *Photo* KD Hertz

It has a cross-section of 1×2 m, and is prestressed with eight cables. The curvature of the prestressing cables is sketched on the photograph. The idea was to counteract moments from uniformly distributed load of a 100 year old masonry facade above, before the 18 m wide opening was made below the beam. Tension is here resisted as unloaded compression in the concrete.

The designer divided the beam into 11 sections in order to cast it stepwise in the old masonry without challenging the stability of the wall. The beam is an example of a straight pearl-chain, where the post-tensioning force presses the sections together and improves the transfer of shear between them in addition to counteracting a moment distribution.

In the case example, the contractor made an error during construction. Over two sections of the beam next to the middle section, he placed the prestressing ducts in a straight line, because he misunderstood the drawing and read numbers for length as numbers for height.

The young engineer had to consider this overnight, because it would be costly and time consuming, if the contractor should remove the two sections and cast them again.

He concluded that this rather serious error did not mean much to the function of the beam, since the resulting lateral force depends on the difference in inclination between the ends and not on the shape of the curve.

Therefore, as long as the curve is smooth and the beam can resist the secondary moments from a less optimal path, the detailed shape is less important.

Another mishap at this particular structure gave rise to good learning as well.

Something hindered pulling one of the cables through its duct. Some cement paste had penetrated one of the duct joints between two sections.

Fig. 6.16 View inside a duct
with penetrating cement.
Photo KD Hertz

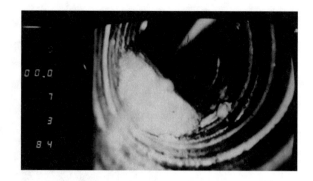

The solution was to call a sewer contractor, who could guide a small rolling robot with a lamp, a camera, and a small drill with a fan into the duct removing the cement paste. Figure 6.16 shows a picture taken by the robot. The whole operation only caused a six hours delay.

In some cases, you may be interested in application of an embedded integrated pearl-chain beam designed to carry the load of adjacent slab elements between supporting columns.

If the integrated transverse part of the pearl-chain beam has the same thickness as the slab, you can avoid the hindrances, which supporting beams usually make for design. For instance with respect to reduction of natural light, increasing storey height, and obstructing building services.

As an example, we can consider an edge-beam integrated across the ends of a number of SL-deck elements. The intention of the beam is to prestress the SL-deck ends across the direction of the span to form a straight pearl-chain beam that can support the deck between columns in a facade.

Consider for example a long office building with SL-decks spanning across the building from facade to facade. We can provide the SL-decks with cable ducts in both ends and create a post-tensioned pearl-chain edge-beam in each facade.

The edge-beam has the purpose of making the position of the columns supporting the deck independent of the element division.

The erection procedure could be as follows: At first, a temporary steel C-profile is mounted on the columns to support the ends of the deck elements.

Then, we pull prestressing cables through the ducts in the elements in the length of the facade and post-tension the deck element ends together as a pearl-chain. Then, we can remove the temporary steel profile, and the edge beams carry the load of the deck structure to the columns.

In the example, we used a span between the columns equal to the width of three deck elements. If we could make the slab elements fixed end supported, their span could be increased 50%. We therefore designed the beam to create a fixed end support.

Likewise, we prestressed the pearl-chain beam to be continuous and fixed end supported over the columns.

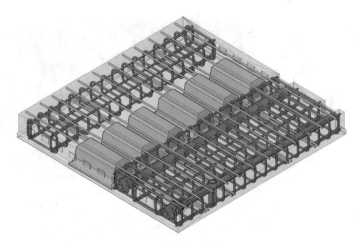

Fig. 6.17 Integrated edge beam test element design [15]

This means that we had to design the pearl-chain edge-beam to resist bending and shear between the columns as fixed end beams and in addition to resist the torsion from the fixed end support of the ends of the SL-deck elements.

A structure like this has been tested for ultimate load in bending, shear, and torsion at the structural lab of the Technical University of Denmark by Iskau and Nissen [15].

Three specimens were tested each of them consisted of three SL-deck elements, which were shortened to be able to fit into the test space. Each deck element was 2.4 m wide and the edge-beam was post-tensioned with two cables.

Figure 6.17 shows the design of one deck element end including reinforcement of the integrated pearl-chain edge beam.

Figure 6.18 shows the test setup, where two hydraulic jacks to the right apply forces to the central of the three deck elements creating a simultaneous bending, shear, and torsion in the edge-beam.

The main conclusion was that the full-scale tests did not show any unforeseen matters, and that you will be able to design a post-tensioned pearl-chain edge beam and utilize its load-bearing capacity by means of common static principles.

6.2.5 Tension Ties

Pearl chains may serve as tension ties, where you can avoid cracks in a tension zone and benefit from the large stiffness of the unloaded concrete section compared to that of the prestressing steel itself.

Tension ties can be separate pearl-chains or imbedded in a number of elements.

Fig. 6.18 Edge beam test setup. *Photo* KD Hertz

For example, you can apply a vertical post-tensioning of both edges of a cross-wall of a high-rise building in order to resist bending from wind load with maximum stiffness and without formation of horizontal cracks as shown in Fig. 6.19a.

Figure 6.19b shows another example.

A continuous beam supports a deck over the ground floor of a theatre. A row of columns supports the beam. However, one of the columns would hinder turning

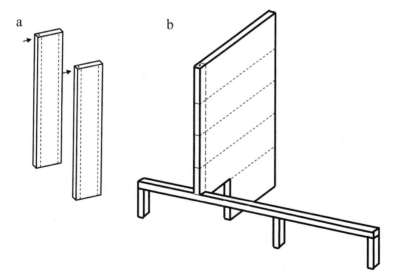

Fig. 6.19 Vertical pearl-chain tension ties **a** shear walls **b** hung up beam

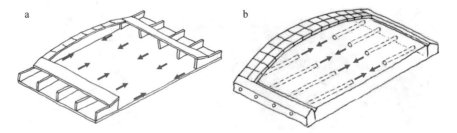

Fig. 6.20 Vault supported by **a** post-tensioned floor slabs **b** drilled tension ties

of thirty metre long carpet roles into a carpet elevator. We solved the problem by designing an integrated tension tie through the edge of the elements of a shear wall above, so that the shear wall could carry the beam in this particular point.

Engineers making arches, vaults, and arch bridges often consider lateral forces to represent a problem that gives rise to design of large foundation blocks. Especially, when you design a flat arch, the lateral forces may become substantial.

In many cases, you can apply the classical solution to this problem by establishing a tension tie between the supports reducing the reaction of the entire structure to be vertical forces.

Designing a building, you may guide the inclined forces of a vault down to the floor slab below. You could do this by supporting the edges of the vault with edge-beams supported by a number of shear walls taking the reactions to the floor as shown in Fig. 6.20a.

You could then post tension the floor slab to take the horizontal components as unloaded compression in the deck elements as a tension tie pearl-chain. In this way, you could create a vault at any floor of a building, even the top floor of high-rise buildings, only resulting in vertical forces on the structures below.

6.2.6 Drilled Tension Ties

In case of a highway bridge, you could sometimes create a similar post-tensioned floor slab as part of the road below. However, if the highway is in service, it will cause a stop of traffic, if you should do this. For a situation like this, you could design a number of drilled tension ties (Fig. 6.20b).

Many contractors are familiar with a technology of drilling a horizontal tube for example with a diameter of 600 mm under a highway while traffic is running.

You could then design a number of these drilled tubes filled with concrete either in-situ cast or as pre-cast segments. You provide each of these horizontal columns with one or more cable ducts and post-tension them.

You have now made a series of horizontal tension ties of prestressed concrete that can resist lateral forces from the arch bridge superstructure as unloaded compression with the benefit of the large stiffness of the concrete cross-section. By means of this, it is possible to build a highway bridge with a minimum of traffic disturbance. Only for a few hours, the traffic must be stopped while the pearl-chain arches are lifted in place.

References

1. Hertz KD (2008) Light-weight load-bearing structures reinforced by core elements made of segments and a method of casting such structures. (Patent on Pearl-Chain technology). Europe EP 08160304.5, USA US 61/080445, PCT (Patent Cooperation Treaty) PCT/EP2009/052987, 13. Patent 2012
2. Freyssinet E (1936) Progrès pratiques des méthodes de traitement méchanique des beétones. (Praktische Weiterentwicklung der Verfahren zur mechanischen Behandlung von Beton, Practical Improvements in the Mechanical Treatment of Concrete), Zweiter Kongress der IVBH = Deuxième Congrès de l'AIPC = Second Congress of IABSE, Berlin-München 1.-11. October 1936. IIb3 pp 205–231
3. Leonhardt F (1964) Prestressed concrete. Design and construction. Ernst W, Sohn, 2nd edn, 677p
4. Brøndum-Nielsen T (1972) Concrete structures II. (In Danish). Structural Research Laboratory, Technical University of Denmark, 125p
5. Brøndum-Nielsen T (1973) Structural concrete. Technical University of Denmark, Structural Research Laboratory, p 136
6. Brøndum-Nielsen T (1976) Partial prestressing. Report R 76, Structural Research Laboratory, Technical University of Denmark, 26p
7. Halding PS (2016) Construction and design of post-tensioned pearl-chain bridges using SL-technology. PhD Thesis. Report R-350, Department of Civil Engineering, Technical University of Denmark, 204p
8. Hertz KD (2007) Light-weight load-bearing structures (patent on super-light structures). Europe EP 07388085.8, USA US 61/004278, PCT (Patent Cooperation Treaty) PCT/EP2008/066013, 21. Patent 2010
9. Hertz KD, Halding PS (2014) Super-light pearl-chain arch vaults. In: Paper nr. 52 in proceedings of the IASS-SLTE 2014 symposium, "shells, membranes and spatial structures", 8p September 2014, Brasilia, Brazil
10. Lund MSM, Hansen KK, Hertz KD (2016) Experimental investigation of different materials in arch bridges with particular focus on pearl-chain bridges. Construct Build Mater 124:922–936. (Elsevier August 2016)
11. Lund MS (2016) Durability of meterials in pearl-chain bridges. Ph.D. thesis. Report R-341, Department of Civil Engineering, Technical University of Denmark, 157p
12. Halding PS, Hertz KD, Viebæk NE, Kennedy B (2015) Assembly and lifting of pearl-chain arches. Paper 71 of fib symposium concrete-innovation and design, 10p. Proceedings p 185. Copenhagen May 2015
13. Hertz KD (1976) The Stringer method applied for calculational design of building structures. (In Danish). MSc project. Department of Building Design, Technical University of Denmark, 147p
14. Hertz KD (1978) Calculation of shear plate structures by means of crossing stringer systems. (In Danish). Bygningsstatiske Meddelelser vol 49, no.4. pp 113–129.
15. Iskau MR, Nissen JS (2016) Design and full-scale test of integrated beam in SL-decks. MSc thesis Department of Civil Engineering, Technical University of Denmark, 256p

Chapter 7
Arch Bridges and Vaults

Abstract Arches and vaults are minimal structures, and basic calculation methods are provided for them. Pearl-chain bridges and vaults are introduced, and a bridge is build based on the technology. The chapter also introduces sandwich arch bridges, which is a method to upgrade traditional arch bridges with increase of the capacity.

7.1 Arches

7.1.1 Arch Principle

Arches are optimal structures capable of carrying uniformly distributed gravity loads in pure compression (without bending moments).

Figure 7.1, shows how pearl-chain structures provide new possibilities for making arches simple, inexpensive, and fast from flat mass-produced deck elements.

As seen from Figs. 1.7, 1.9, and 1.10, builders have applied arches for thousands of years for vaults in buildings and for arch bridges. Some of these arch structures were in service for millennia.

In general, the advantage of arches is that you can carry a uniform load by a minimal structure made of inexpensive and sustainable materials, which are mainly suitable for taking compression such as stone, mortar, and concrete. Furthermore, you can avoid the expensive and less sustainable materials to resist tension.

Figure 7.2 shows a comparison between a possible force distribution in an arch and in a beam with fixed ends, uniformly distributed load p and span L. Both get the same moments if they are simply supported. A beam with fixed end supports get reduced moments, but the difference between maximum and minimum moments is unchanged. The arch can totally avoid moments for a uniformly distributed load by fixing the supports for lateral movements. You find the lateral arch force from a condition of no moment at the midpoint.

If you have live load on a bridge or a vault, you cannot avoid bending moments.

A single load P is often critical in a quarter point of the span at L/4 from the one end at a height h. It gives vertical reactions of P/4 and 3P/4. If you consider your arch to have a hinge in the middle, you can take moment about that, and you get a

K. D. Hertz and P. Halding., *Sustainable Light Concrete Structures*, Springer Tracts in Civil Engineering, https://doi.org/10.1007/978-3-030-80500-5_7

Fig. 7.1 Pearl-chain arch bridge across Vorgod River, Denmark. *Photo* Abeo Ltd.

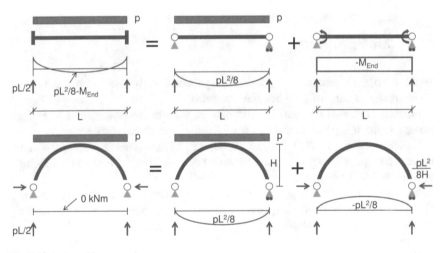

Fig. 7.2 Arch and beam statics

lateral reaction force of size PL/8H and the moments in the quarter points at height h are then PL(3/16 – h/8H) and PL(1/16 – h/8H).

However, dead load is often a relatively large part of the total loading. The large normal force in the arch gives you compressive stresses, which in traditional stone arches allows adoption of tensile stresses from secondary moments as an unloading of the compression.

7.1.2 Arch Shapes

Which curve is then the optimal for the dead load? This depends on the load on the arch. If you have a vertical load, which is uniformly distributed over the span, the optimum is a parabola. If the arch must only carry its own weight, the optimum is a chain line also called a catenary. If you have both, the optimum can be a circle for flat arches, because it will be close to the average of the other two shapes.

If the arch span is for example eight times its height, the ratio between the height in the quarter point and the arch height at mid-span will be 0.750 for the parabola, 0.765 for the catenary, and 0.761 for the circle and the curves looks as shown in Fig. 7.3 at the top.

When the arch span is four times its height, the ratio between the height in the quarter point and the arch height at mid-span will still be 0.750 for the parabola, 0.765 for the catenary, and now 0.791 for the circle and the curves looks as shown in Fig. 7.3 at the bottom.

If you make arches and vaults with circular curves, you have the big advantage of being able to build them from equal elements. As you see, the circular arch is in general the most optimal curve to apply for relatively flat structures.

Builders have therefore applied circular arches to support walls over openings for windows, doors, and gates and to make daring bridges (Fig. 1.10).

By increasing the width of the arches, you get circular vaults suitable as roof structures (Fig. 1.9) or, ceilings in large rooms supporting the upper stories.

In tunnels and sewers, builders have used circular vaults to sustain the earth pressure and in many cases, they have made the cross-section a full circle since their structure should not only receive forces from above, but also deliver them to the underlying material and sustain pressure from the sides.

Designers have not only applied circular arches and vaults to sustain vertical forces, they have also e.g. made horizontal compression rings around holes in the top

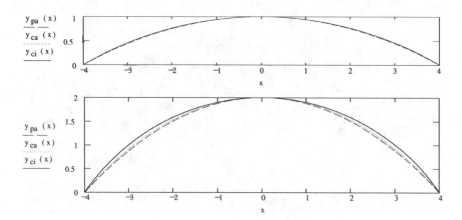

Fig. 7.3 Arch curves: Parabola y_{pa}, catenary y_{ca}, and circle y_{ci} for H/L $= 1/8$ and $1/4$

of cupolas (Fig. 1.14), and horizontal arches to support distributed lateral pressure from torus shells as in Fig. 1.11, from soil as in Fig. 1.12, or from water in dams.

7.2 Pearl-Chain Arches

7.2.1 Pearl-Chain Arch Bridges

Builders have always had the main disadvantage that construction of a compression arch required a temporary scaffolding resisting the loads until the arch was completed.

The pearl-chain principle allows you to build arches without scaffolding and thereby make arches more applicable. You can especially make a circular pearl-chain arch simple, inexpensive, and fast, because you can design it from equal, flat, and mass-produced elements.

In order to avoid all expensive curved supports while assembling pearl-chain arches, the authors have found that a successful method is to place the elements in a vertical position on a plane surface before post-tensioning them together [1, 2]

In Fig. 7.4, a pearl-chain arch is lifted by crane after the assembly.

Prior to the post-tensioning, the joints of width 2.4 m were cast out in the vertical position. The method for pouring the joints were tested in 2015. It was proven that a mortar could be used to fill out the joints sufficiently [3, 4].

The vertical assembling method allows a simple procedure of erection.

Flat SL-deck elements are transported to a place near the building site, which for a freeway bridge may be at the roadside.

Fig. 7.4 Pearl-chain arch after assembling. *Photo* KD Hertz

Fig. 7.5 Full-scale test of pearl-chain bridge for skew load. *Photo* KD Hertz

Here, the contractor assembles all arches and casts out the joints before a specialized crew arrive to prestress the pearl-chain structures with post tension cables.

After that, you can lift all arches in place quickly without disturbing the traffic more than about 4–5 h for a typical freeway bridge.

Before the first pearl-chain arch bridge was established, we made a full-scale test for skew load. It was a concentrated load in one of the quarter points, which as explained in 7.1.1 traditionally is the most dangerous load case for an arch bridge.

Figure 7.5 shows the test setup. It proved to sustain almost the double load, for which we had conservatively designed it [5–7].

Figure 7.6 shows a similar test made in 1893 before a Monier arch was applied by Asger Ostenfeld in the Gefion Bridge in Copenhagen in 1894 (Fig. 7.7).

The bridge stood 104 years until the community would like to change its position.

7.2.2 Long Span Pearl-Chain Arch Bridges

Sometimes engineers would like to build an arch bridge with a long span.

Of course, they would like to do that without application of a big costly curved scaffolding.

Gustave Eiffel gave a solution to this problem. In 1877, he constructed the Maria Pia Railway Bridge across the Douro River in Porto with a single steel lattice arch spanning 160 m and with a height of 60 m.

Figure 7.8 shows his solution. He prolonged the two towers next to the basements of the main arch with temporary towers between which he made a cable like that of a suspension bridge. He applied this cable to lift parts in place and combined with

Fig. 7.6 Monier arch test for skew load 1893 before building the Gefion Bridge

Fig. 7.7 Gefion Bridge with Monier arch from 1894. *Photo* KD Hertz

inclined cable stays, he could support parts of the arch before it was completed and could carry itself.

The same procedure could be applied for construction of a pearl-chain arch with a long span. The arch could be constructed by two half pearl-chain arches or of smaller segments kept in place and temporary supported by a main cable between temporary towers and by inclined cables to the sides.

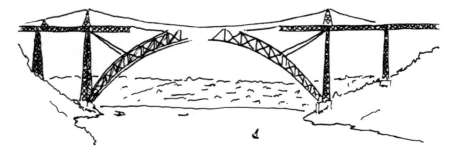

Fig. 7.8 Maria Pia Bridge. Drawing KD Hertz

For small flat arch bridges, you can transfer the distributed load from the road or rail to the arch by means of a filling material. For large arch bridges, you may have to apply intermediate columns. The columns act with concentrated loads on single points of the arch.

This means that you do not need the curvature of the arch in between the points, where it supports the columns for anything else but the dead load of the arch itself.

You can therefore often make a more optimal pearl-chain arch using straight elements connected at angles at the points, where it supports the columns. You may perhaps include special elements at the support points providing the angles and supporting the columns.

7.2.3 Half-Arch Structures

In many cases, you can take lateral forces from arches and vaults in an even more simple way by application of half-arches. We originally developed the method for bridges crossing streams, where tension ties between the supports would be impossible, and where soft soil meant that pile foundations could not resist lateral forces.

At each side, you supplement the arch with a half-arch that will provide the same lateral pressure as the main arch. See Fig. 7.9. You counteract the normal force of the two half arches by a tension force in a tie above the arches. The tie may be a post-tensioned deck capable of resisting tension as unloaded compression.

You can consider the structure as consisting of four half arches or of two sections with cantilevered half arches to each side counterbalancing each other.

A series of these types of super-light structures were already tested to failure by Larsen [8] before the pearl-chain principle was invented (Fig. 7.10).

Two half arches were cast in a curved mould and embedded in light aggregate concrete. A tension tie in the top transferred its force to the compression arch with a steel plate at each end. At failure load, the strong arch crushed, where its end penetrated from the light concrete and was loaded by the steel plate (Fig. 7.11).

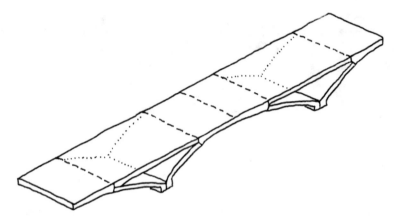

Fig. 7.9 Half-arch solution with horizontal tension tie

Fig. 7.10 Test of super-light double half arch. *Photo* KD Hertz

The half-arch structure is suitable for creating large vaulted spaces at any storey in a building, because it only transfers vertical loads to the lower levels. The outer half arches allow architects to design open facades without columns. This is possible, since a cantilevered half arch can carry vertical loads.

Half-arch structures are also suitable for designing bridges over streams, roads, and rails, where the outer half arches fit to the sloping embankments that you often find to the sides of the highway lanes or rail tracks. In Sect. 7.3.4 it is shown, how you may shorten the outer half arches if required.

Fig. 7.11 Fracture of the end of a strong arch core. *Photo* KD Hertz

7.3 Pearl-Chain Sandwich Arches and Vaults

7.3.1 Pearl-Chain Sandwich Arches

Traditional arch builders have faced two main problems.

(1) A requirement for costly curved scaffolding.
(2) A requirement for heavy foundations to resist the lateral forces.
(3) A requirement for a an arch cross-section that can resist secondary moments.

The first problem is solved by application of pearl-chains introduced in Chapter 6 and in Sect. 7.2.

The second problem can be solved by application of post-tensioned ties as shown in 6.2.5 and in 6.2.6 and by application of half arches as shown in 7.2.3.

The third problem can be solved by application of what we call sandwich arches.

Skew loading causes the secondary moments, typically from a concentrated load close to the quarter point.

Figure 7.12 shows Ponte de Mosteiró constructed by Prof. Edgar Cardoso in 1972 across the Douro River. It has a span of 110 m and its arch height is only 7.2 m. It is an example of a half arch structure, as introduced in 7.2.3.

It is also an example of a bridge, where a lattice—here an X lattice—is capable of transferring shear between arch and top slab, so that a pair of forces can resist secondary moments from live load with the entire lattice height as lever arm.

Eugène Freyssinet applied this trick first in the Boutiron Bridge from 1912 across the river Allier near Vichy. In this bridge he applied a V-lattice to transfer the shear and he established a span 72 m with a height of 5.2 m.

Fig. 7.12   Ponte de Mosteiró. *Photo* KD Hertz

This appears to represent a general solution to the old problem of secondary moments from skew loads on arch bridges. For smaller arches, it will not pay back to build a lattice. Here, the obvious solution is to establish a layer of material that can transfer shear between the arch and a top plate as shown in Fig. 7.13. The top plate can serve as basis for a rail or a road pavement, or can function as the pavement.

The cast shear transferring material could for example be a light pervious concrete with a modest tensile strength of a few MPa. Such a pervious concrete could ensure that penetrating rain is guided to the top of the arch and away from the bridge.

For positive moments with compression at the top, you may then resist the tension by unloading the compression force in the arch.

In case you would avoid cracks in the top plate for negative moments, you could decide to post-tension the top plate after it has been cast in situ or placed as a layer of concrete elements.

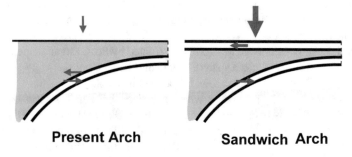

Fig. 7.13   Principle of a sandwich arch

Fig. 7.14 Test of a scaled sandwich arch bridge

Figure 7.14 shows one of a series of model tests made with the sandwich arch principle in order to unveil, if anything should hinder a solution like that.

Nanna Tange Bech and Maria Holst [9] created a series of scale model SL-deck elements in strong concrete and with small blocks of model light aggregate concrete. They then prestressed these model SL-deck elements to a number of pearl-chain arches.

A light filling was then poured on top of the arch, and they finished by making a series of scaled massive deck elements and prestressed them to become pearl-chain top plates.

They tested the arches with varying stiffness of the filling between arch and top plate.

The tests were performed with a point load in the quarter point.

The scaled tests demonstrated that the sandwich arch principle can increase the load-bearing capacity with respect to secondary moments from skew load more than ten times.

Figures 7.15 and 7.16 illustrates a simple version of the sandwich arch, where a pervious light concrete is cast out between a pearl-chain arch and an in-situ cast top slab. Here, the sandwich layer has simple inclined sides. This was applied in the project shown in Fig. 7.15 for a pedestrian bridge designed by Henning Larsen Architects Ltd from 2015.

7.3.2 Sandwich Arch Bridges

Figure 7.17 shows a bridge designed and built in 2015 across Vorgod River by Sweco Engineers Ltd and Abeo Ltd. It has a main arch with a span of 13 m and two half arches of 6.5 m each.

Fig. 7.15 Pearl-chain Sandwich arch bridge project. Henning Larsen Architects Ltd.

Fig. 7.16 Simple sandwich arch principle

Fig. 7.17 Super-light pearl-chain bridge across Vorgod River

As described in Sect. 7.2.3 the half-arch structure is capable of lifting the lateral forces from the main arch up to be withstood as tension or unloaded compression in the bridge deck.

Here it is required, and it is a convenient way of building, because the soil is soft and the bridge is supported on a pile foundation that cannot give sideward resistance.

Figure 7.1 shows the same bridge during construction. At first, the contractor established the pile foundations. Then, he made a preliminary sideward stabilisation of them by means of horizontal prestressed Macalloy rods as tension ties.

Next, he assembled the pearl-chain arches and lifted them in place, and the lifting operation took only five hours.

This demonstrates that you can reduce interruption of traffic to this time, when you establish a pearl-chain bridge across a freeway.

With the arches in place, a pervious concrete was cast on top of them. It was developed by Lund [3] in collaboration with researchers at Iowa State University to be cast in sufficient thicknesses for the purpose. At last, the top slab was cast, and the preliminary tension ties could be removed.

7.3.3 Cassette Sandwich Arches

We designed the arch in the Vorgod River bridge in Fig. 7.17 as a pearl-chain of flat SL-deck elements. The SL-deck factory could mass-produce them quickly and inexpensively as a part of their ongoing deck element production.

However, for an arch in a bridge you do not need all advantages of the SL-decks such as sound insulation. In many cases, you can therefore save material, money, and CO_2 by application of what we call cassette elements similar to the one introduced in Fig. 6.12 in Sect. 6.2.2.

A cassette consists of a frame of strong concrete filled with a plate of light concrete in less height. The outer sides of the frame serve as sides of the joints to the neighbour elements. In the longitudinal direction, the sides are inclined in order to give the curvature and shape of the arch. The longitudinal sides of the frame contain tubes for post-tension cables.

Using such cassette elements, you can benefit from the voids for more than reducing weight. If you turn them upwards as shown in Fig. 7.18, the shear transfer-ring filling material can get an effective grip in the top surface of the arch. That will ensure that there is no slipping between the surface of the arch and the filling.

It is also possible to apply the sandwich principle to ensure lateral stability of structures consisting of a number of high and slender arches.

In Fig. 7.19 is shown such an example of two adjacent long and slender arches.

It is Ponte da Arrábida, which Edgar Cardoso constructed in 1963 in Porto, Portugal.

It has a span of 270 m, a height of 70 m and a width of 27 m.

Fig. 7.18   Reversed ribbed arch of cassette elements

Fig. 7.19   Ponte da Arrábida. *Photo* KD Hertz

In order to give them stability for transverse forces, he constructed the arches with a certain distance and connected them with a horizontal shear transferring X-lattice.

The shear transfer hinders that the arches can move relative to each other in their longitudinal direction, as they would, if they should tilt sidewards.

Likewise, you may design material saving and slender pearl-chain arches separated by light concrete sections serving as shear plates.

A special design of such pearl-chain arches for bridges and vaults is by application of the ribbed cassette elements as shown in Fig. 7.18. Here the sides of the cassettes constitute the arches, which are separated by the light concrete in the shear plates.

This can create very stable arch structures for bridges and vaults with a minimum of materials applied for the elements.

7.3.4 Lattice Effects in Sandwich Arches

If you want to improve the shear capacity of the filling material of a sandwich arch or perhaps even apply a material that can only resist compression, you could introduce some vertical or inclined tensile members. For example, such members could be made as pre-tensioned columns of strong concrete, which can withstand tension as unloaded compression. They can then transfer shear as tension in the columns and compression in the filling material as shown at the top of Fig. 7.20.

The figure also shows examples of how you may save filling material in deep sandwich arches by placing horizontal round or triangular tubes creating cavities or sand filled volumes around which the filling constitutes a lattice. As shown in Fig. 1.10, builders have applied such holes since antiquity to save materials and to unload sidewards pressure from streaming water, when the water level of the river is high.

Finally, the figure shows how you may reduce the length of outer half arches and still counterbalance the horizontal arch force by introducing a vertical tension force for example from a soil anchor like those applied for sheet piling or by a concrete block or a box with sand.

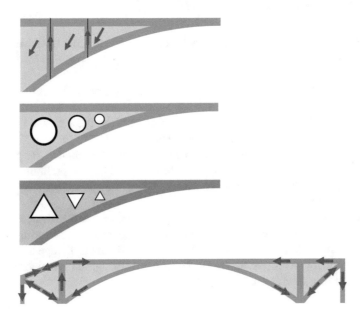

Fig. 7.20 Lattice formation in sandwich arches

Fig. 7.21 Gallery bridge project. Project and rendering by Timo Nielsen et al. 10

7.3.5 Shear Wall Sandwich Arches

Figure 7.21 shows a project for a gallery bridge intended to cross the River Thames in London by Nielsen et al. [10]. It has a main arch with a span of 176 m and two half arches each spanning 88 m.

In this project, the sandwich effect is established by means of shear transferring walls.

The walls make it possible to utilize the space between the arch and the deck of the bridge as a two storey high gallery building.

For the construction is was suggested to establish a preliminary support at the middle of the main span. An alternative solution could be two preliminary pylons over the main foundations in the river with tension cables, as suggested by Gustave Eiffel and shown in Fig. 7.8.

At the centre span, the project group designed a top arch that extended the space for the gallery and provided the bridge with a supplementary load-bearing capacity.

Such a double effect is also seen in the Simone de Beauvoir Bridge in Paris from 2004 by Dietmar Feichtinger, and in the Royal Albert Bridge at Plymouth from 1854 by Isambard Kingdom Brunel.

References

1. Halding PS, Hertz KD, Viebæk NE, Kennedy B (2015) Assembly and lifting of pearl-chain arches. In: Paper 71 of fib symposium concrete-innovation and design, 10p. Proceedings p 185. Copenhagen May 2015

2. Halding PS, Hertz KD, Schmidt JW (2015) Precast pearl-chain concrete arch bridges. Eng Struct 103:214–227. (Elsevier 2015)
3. Lund MS (2016) Durability of meterials in pearl-chain bridges. Ph.D. thesis. Report R-341, Department of Civil Engineering, Technical University of Denmark. 157p
4. Lund MSM, Hansen KK, Hertz KD (2016) Experimental investigation of different materials in arch bridges with particular focus on Pearl-Chain Bridges. Construct Build Mater 124:922–936. (Elsevier August 2016)
5. Halding PS (2016) Construction and design of post-tensioned pearl-chain bridges using SL-technology. Ph.D. Thesis. Report R-350, Department of Civil Engineering, Technical University of Denmark, 204p
6. Halding PS, Hertz KD, Schmidt JW, Kennedy B (2017) Full-scale load tests of pearl-chain arches. Eng Struct 131:101–114
7. Rossi M (2015) Pearl-chain arch bridges above electric railways. MSc thesis, Universitá degli studi di Genova, 132p
8. Larsen F (2007) Light concrete structures. MSc project, Department of Civil Engineering, Technical University of Denmark, 117 p
9. Bech NT, Pedersen MH (2015) Development and test of arch bridge with materials of different stiffness. MSc thesis, Department of Civil Engineering, Technical University of Denmark, 235p
10. Nielsen T, Henriksen T, Bendtsen R, Østerby I, Petersen F (2015) Gallery bridge. Project in super-light structures. Technical University of Denmark, 31p

Chapter 8
Shells

Abstract Element cupolas are possible to create by using light trapezoidal concrete elements and pearl-chain reinforcement. A calculation method is provided for cupolas, and other possible modular shell shapes are introduced together with tubes, tunnels, and floating Super-light structures.

8.1 Shell Structures

8.1.1 Shell Structures

Shell structures can give powerful architectural expressions and make it possible for builders to create large spaces in buildings. You can make shells with application of a minimum of building materials and shells therefore have a huge potential for sustainable building design (Fig. 8.1).

Unfortunately, architects and engineers do almost not apply shell structures in modern buildings. This is mainly because shells usually require costly manmade, customized moulds, complicated casting, and demanding engineering calculations.

However, the technologies dealt with in this book can remove all these problems.

We therefore hope that the great potential of shells will be applied more often to create functional and expressive structures (Fig. 8.2).

8.1.2 New Possibilities

Pearl-chain technology makes it possible to create inexpensive curved shapes by straight, mass-produced elements, which can easily be transported and lifted in place. See Chap. 6.

Super-light mass-produced cassette elements as shown in Fig. 6.12, opens possibilities for creating light curved shell structures, which are inexpensive and gives a small material consume, -building cost, and -CO_2 impact. You can apply textile

Fig. 8.1 Polylithic. A super-light modular shell structure by Peters et al. [1]

Fig. 8.2 Super-light tube bridge by Herdal et al. [2]

moulds due to the small mould pressure of light concrete, and they open new potentials for creating shells with double curved shapes (Sect. 8.6.2). Inflatable moulds
enables builders to create double curved structures and cassette elements with vaulted
fillings of light concrete inside frames of ordinary load-carrying concrete, as for
example shown in Figs. 6.11 and 7.18.

In addition to vaults, which we mainly deal with in Chap. 7, the present chapter
presents:

- Cupola design with only a few different elements and built without scaffolding,
- Torus structures for round buildings with open facades.
- Modular elements for creating free shapes of shells,
- Hyperbolic shells,
- Tubular structures and tunnels,
- Floating submerged tunnels,
- Structures cast underwater,
- Floating bridge foundations,
- Ships of concrete.

8.2 Cupolas

8.2.1 Cupolas

Builders have used concrete cupolas or dome structures for thousands of years (Figs. 1.14 and 3.3) to create large indoor spaces. A cupola is a statically indeterminate structure, which according to Sect. 3.2 means that you as a builder can apply direct engineering and determine where the forces should be.

For dead load, which usually is the largest load contribution, and for uniformly distributed live load, it would be most natural to apply a symmetrical set of compression forces along the medians. For a flat cupola as the one shown to the left in Fig. 8.3, equilibrium of the median compression forces and the weight of the elements will give rise to compressive horizontal ring forces, except at the bottom, where you need to resist outwards horizontal components of the median compression forces.

You can resist the reaction forces either by a solid foundation or by a ring shaped tension tie that you can create as a prestressed pearl-chain concrete ring withstanding the tension as an unloading of the pre-compression.

Piere Luigi Nervi used a post-tensioned ring foundation in his Palazzetto dello Sport as seen in Fig. 3.3.

At a certain inclination of the cupola, it is not considered flat any more, and ring tension forces occur in the shell as well. At this specific inclination, Nervi introduced

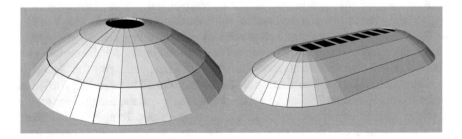

Fig. 8.3 Cupola and hall

glass windows in the Palazzetto and guided the forces along the radians into inclined columns supported on the prestressed ring foundation. The same is done in the cupola of Hagia Sophia in Istanbul.

8.2.2 Element Cupolas

You can divide a cupola in halves and insert a cylindrical vault (Chap. 7) as shown to the right in Fig. 8.3. You may then transfer the horizontal compression- or tension forces of the two half cupolas by horizontal stringers in the vault. You may transfer the horizontal lateral forces of the vault itself as unloaded compression in a post-tensioned floor slab or in a series of post-tensioned concrete rods cast in tubes under the floor as explained in more detail in Sects. 6.2.5 and 6.2.6.

You can insert windows at the top of the vault by means of the earlier described cassette elements, where you can exchange the light concrete filling with glass. You can apply the strong cassette edge profiles as "columns" to transfer compression forces of the vault in the direction of the arch and as "beams" to transfer compression forces from the one half cupola to the other in the longitudinal direction.

You can make a cupola of plane mass-produced prefabricated elements. To understand how, you may first consider the well-known construction method of prefabricated slurry tanks. Factories produce such tanks ready-made with diameters up to about 60 m.

First, you make a ring-shaped foundation. On that, you place a number of equal plane rectangular concrete panels vertically, with a small angle between them and then grout the joints. Then you pull circumferential post-tensioning strands around the tank and tension them with a device such as a DYWIDAG ME Floating Coupler, where the strands overlap and you stress them by means of a simple belt-buckle principle.

This means that no anchoring blocks and complicated anchoring reinforcement is required.

The ring-strands prestress the circle of concrete panels so that they can resist tension from the pressure of the content of the tank as unloaded compression with the stiffness of the concrete and without cracking.

Now imagine that you instead of vertical rectangular panels place a circle of trapezoidal panels with an inclination inwards to the centre of the circle. The geometry of the elements will determine how much they tilt, since they should lean against each other as seen in Fig. 8.4. The contact between the inclined elements will give rise to a lateral compression at the top and tension at the bottom ring foundation.

The ring foundation should therefore be post-tensioned to resist this tension as unloaded compression. The lateral compression means that the tilting elements are stable as long as they have adjacent elements on both sides. To give the first tilting panels an extra support during construction, you may place a pole at the centre from which wires can fix upper corners of the panels as also shown in Fig. 8.4. Strictly speaking, you only need a preliminary support for the first element. When you have

Fig. 8.4 Building a modular cupola

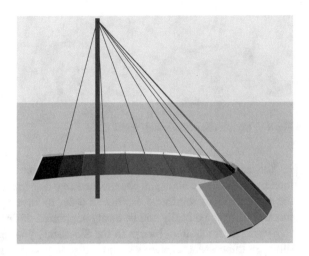

established the first full ring of plane trapezoidal elements and grouted the joints between them, you can repeat the process with a second and a third ring. The result is a cupola as seen in cross-section at Fig. 8.5. It consists of only three different mass-produced plane trapezoidal concrete panels. It has a hole at the top that you may cover with a panel without any structural importance or leave it as an opening or a window.

It would often be beneficial to design the trapezoidal panels as cassette elements as shown in Fig. 6.12. The outer frames of strong concrete then transfer the compression forces along the medians (the longitude lines) and the lateral compression forces along horizontal rings.

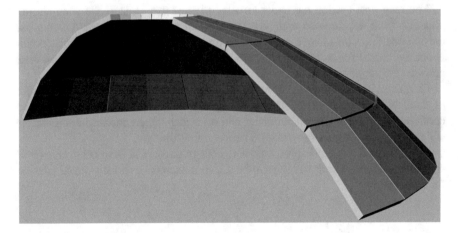

Fig. 8.5 Cross-section of a cupola of three elements

Fig. 8.6 Construction of element cupola [3]

You can make the cross-section of the frames more slender in the upper layers and thereby reduce the weight and change the force distribution. Usually you would keep the same depth of the frame profiles in order to have a joint between the elements that can be grouted easily and be easily supported during erection.

Similar to the bottom blocks in the SL-decks, you may leave the light concrete filling in the cassette elements without cladding and use it for acoustical damping. A light aggregate concrete may give the same sound damping as a wood wool panel that you often apply for this purpose alone. See more about this in Chap. 4. This feature is very relevant, since a cupola acts as a concave acoustical mirror with an unpleasant concentration of reflected sound.

The only tension reinforcement required is the pretension wires at the circular foundation that you could establish as a circular pearl-chain of equal straight prefabricated elements. The foundation elements could be produced with a length equal to the bottom width of the above trapezoidal panels. The only scaffolding applied is a pole at the centre or perhaps just a rod supporting a single element in each layer.

You may support the cupola by inclined arches as shown by Ottersland [3] in Fig. 8.6. Four panels, each made as a pearl-chain arch, create a basis to support the cupola.

You may post tension the upper circular edge of this basis, if the compression obtained from the inclined arch structures is not sufficient for counteracting the tension that the cupola above can give. See more about this in 8.2.3. You may counteract the lateral forces from the arches by prestressing the floor slab in the lines between the supporting points.

Calatayud [4] shows in Figs. 8.7 and 8.8 how to design an elegant cupola structure of trapezoidal cassette elements. She designs it with a triangular floor plan and combines two of them into a parallelogram floor plan, where inclined arches at the edge panels allow a void for windows between the cupolas.

The structural principles presented here change the cupola from being an expensive and time-consuming structure to be a fast, inexpensive and sustainable design for large spaces.

Fig. 8.7 Triangular cupolas
[4]

Fig. 8.8 Triangular cassette element cupolas [4]

8.2.3 *Cupola Design*

When you calculate the forces in the cupola, you start from the top, where you have no upper loads. Figure 8.9 shows the situation, when you reach layer number i of height H_i with an upper horizontal radius r_i and lower radius r_{i+1}.

Here you consider one of the trapezoidal elements with a load P_i at the horizontal distance d_i from the bottom edge. You already know the accumulated vertical load P on this element from the upper layers that will act at the distance $r_{i+1} - r_i$ from the bottom edge.

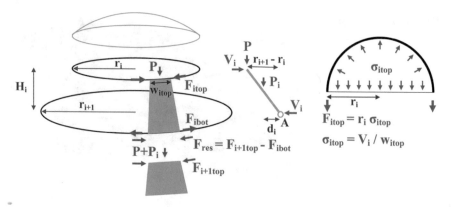

Fig. 8.9 Force distribution in a cupola of trapezoidal panels

You now find the moment of these two loads about the bottom edge shown as Point A in the middle of Fig. 8.9.

$$M_i = P(r_{i+1} - r_i) + P_i d_i$$

From that, you determine the horizontal forces V_i required to counterbalance this moment and the force per meter at top and bottom. If you consider the top width w_{itop}, you find this force per meter σ_{itop} from

$$V_i = M_i/H_i, \; \sigma_{itop} = V_i/w_{itop}$$

This force per meter represents a circumferential compression force

$$F_{itop} = r_i \sigma_{itop}$$

It may be easier to comprehend this by considering the horizontal load as a uniformly distributed "hydrostatic" pressure σ_{itop} in a slice of a cylinder as indicated to the right at Fig. 8.9. This pressure acts on the wall of the cylinder, and it acts in the cross-section of the cylinder in the figure. The force F_{itop} must therefore be in equilibrium with the pressure over a length corresponding to the radius r_i.

Now you do the same in order to find the circumferential tension force F_{ibot} at the bottom of the layer number i and the compressive force F_{i+1top} at the top of the layer i + 1 below. The difference between these forces gives the resulting circumferential compression force at the bottom of layer i as

$$F_{res} = F_{i+1top} - F_{ibot}.$$

For flat cupolas, you will get compressive circumferential forces at all levels except at the bottom, where you as mentioned have to prestress the ring foundation to resist the tension force.

In case you design a cupola, you will get tensile forces in the shell, and you can provide the bottom of the trapezoidal elements with ducts or grooves for post tensioning a ring reinforcement at the layers, where it is relevant.

8.2.4 Cupola Element Design Example

Consider an element cupola with a diameter $D_{cupola} = 30$ m, and a hole in the top with diameter $D_{hole} = 5$ m. The total height is $h_{cupola} = 10$ m. The cupola has a shape as a spherical cap. It consists of flat trapezoidal elements in three layers, with 18 elements in each layer. All elements have the same length, but the width will decrease, as we get closer to the top hole.

We wish to find the size of the elements.

Due to the spherical shape, any cross section through the centre of the cupola will be circular with diameter $r = 16.25$ m, and the equation of the circle part is

$$y(x) = h_{cupola} - \left(r - \sqrt{r^2 - x^2}\right)$$

The arch length s is found by determining the cupola start angle v (Fig. 8.10)

$$v = \arcsin\left(\frac{D_{cupola}}{2 \cdot r}\right) = 67.4^{\circ}$$

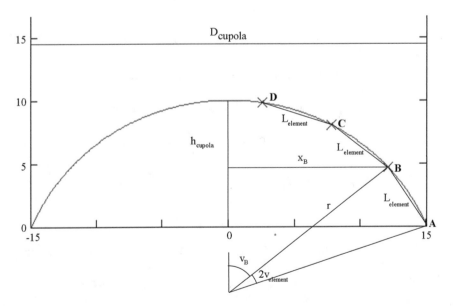

Fig. 8.10 Cupola cross section

$$s = 2 \cdot v \cdot r = 38.2\,\mathrm{m}$$

The cross section arch length is subtracted with the arch length of the top hole, and divided with the six concrete elements in the cross section to find the arch length of each element $s_{element} = 5.53$ m. Reusing the above equations, we can find the straight element length $L_{element}$ when knowing the half angle between the radius r to the ends of each element, $v_{element} = s_{element}/(2\,r) = 9.76$ deg:

$$L_{element} = 2 \cdot r \cdot \sin(v_{element}) = 5.51\,\mathrm{m}$$

To determine the width of the elements in each of the four levels of the cupola (**A, B, C** and **D**), it is necessary to find the diameter of the spherical cap at each level. For instance, the angle from vertical to the point **B** is $v_B = v - 2v_{element} = 47.87$ deg. The x-coordinate of point **B** is then $x_B = r\sin(v_B) = 12.05$ m.

This is also the radius of the spherical cap at the level of the connection between bottom- and middle element, and hence, the circumference length is

$$O_b = 2\pi \cdot x_B = 75.72\,\mathrm{m}$$

Now, the circumference length is divided into the 18 elements of equal widths $s_B = O_B/18 = 4.21$ m, and the straight width w_B (flat elements) is determined from the start angle of the circle parts

$$w_{elB} = 2 \cdot x_B \cdot \sin\left(2 \cdot \frac{\pi}{2 \cdot 18}\right) = 4.19\,\mathrm{m}$$

The same method can be used for the remaining levels, and the final widths of elements in each level are:

- Level **A** (cupola bottom): $w_{elA} = D_{cupola} \cdot \sin\left(2 \cdot \frac{\pi}{2 \cdot 18}\right) = 5.21$ m
- Level **B**: $w_{elB} = 4.19\ m$
- Level **C**: $v_c = v_B - 2v_{element} = 28.36\ deg, \quad x_c = r \cdot \sin(v_c)$,

$$w_{elC} = 2 \cdot x_c \cdot \sin\left(2 \cdot \frac{\pi}{2 \cdot 18}\right) = 2.68\,\mathrm{m}$$

- Level **D** (edge of hole in top): $v_D = v_C - 2v_{element} = 8.85\ deg, \quad x_D = r \cdot \sin(v_D)$,

$$w_{elD} = 2 \cdot x_D \cdot \sin\left(2 \cdot \frac{\pi}{2 \cdot 18}\right) = 0.87\,\mathrm{m}$$

The flat elements can now be designed, for instance as cassette elements (similar to Fig. 6.12), where there can be positioned post-tensioning cables around the circumference in levels where it is required, to avoid cracking from the ring forces of the cupola. The ring forces can be found as explained in 8.2.3, when the element geometry and therefore the dead load is known.

8.2.5 Half Cupolas

As seen from Fig. 1.13, half cupolas have also been popular structures for thousands of years.

The force distribution of a half cupola is more complicated than for a full.

However, a rather simple approach for direct engineering (see Sect. 3.2) may be to design it as explained above for the full cupola and introduce an arch of strong elements at the edge of the opening as a curved beam with lateral load.

The elements at the curved edge beam are subjected to shear and moments from the horizontal forces. These edge beam elements therefore probably need to be massive and made of strong reinforced concrete.

Half cupolas may have a number of applications. They may for example serve as pleasant cool shelters in landscapes. They can also be used to accommodate theater performances and concerts for an open-air audience by creating a shelter for the artists and at the same time constitute an acoustical reflector.

By choosing the quality and type of materials for parts of the inner surface of the half cupola, you can support sound reflection as a function of frequency so that you for example can damp the deep tones and support reflection of the high tones, which often are reduced the most on their way to the audience. See more about acoustical damping in Sect. 4.1.5.

8.2.6 Torus Part Shells

You can design torus part shells by means of a technique similar to the one used for the cupolas. Again, you may apply for example only three trapezoidal elements and again cassette elements could be advantageous. Genius master builders like Apolodorus of Damascus made torus part shells as seen from Fig. 1.11 where he adopted the lateral forces as compression in a horizontal ring beam supported by columns to the left in the photograph.

It is possible to combine a cupola and a torus part shell as shown in Fig. 8.11. A vertical cylindrical wall or tube at the edge of the cupola supports both. The horizontal forces from the torus part are resisted by a circumferential prestressed cable at the outer edge. All other parts are subjected to compression, and you may design the structure by the same principles as shown for the cupolas by means of moment equilibrium about a lover edge of each element level.

As also indicated in Fig. 8.11 in the top drawing, you can apply a filling for example with a light concrete, at the top of the torus and cupola, if you want to apply the structure as floors in for example a round tower. You can supplement this with a circular concrete slab that can assist carrying the load by resisting tension in the radial direction. If the filling can transfer shear, the torus will act as a sandwich arch with increased moment capacity. The slab can be prestressed by the compression provided by tensioning the circumferential prestressing cable, so that tension will only unload

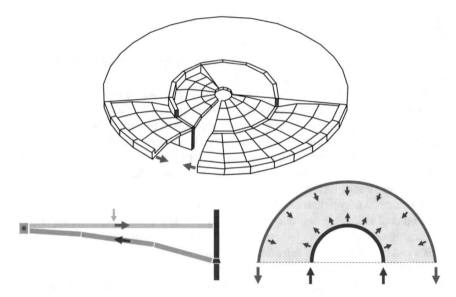

Fig. 8.11 Torus part shell and cupola supported by a cylindrical wall

the compression, and the large stiffness of the concrete is therefore utilized. A special application for a structure like this is reutilization of silos.

8.3   Modular Shells

8.3.1   Modular Shell Structures

Above, we have dealt with vaults, cupolas and torus part shells. However, the shell concept comprises infinitely many structures of which a good share can benefit from the technologies of light and super-light concrete. Designers can make curved strong spines of simple mass-produced concrete parts as load-bearing parts.

They can then fill out the space between the spines with light concrete cast against double curved textile moulds. The moulds can be attached to the spines and perhaps be made by inflated air or provided partly with inflatable air channels and stiffened with rods in order to create the desired curvature.

These technologies already exist and require only few resources. They remove the main practical hindrances for application of shells for in-situ cast structures as well as for shell structures assembled from modular prefabricated shell elements.

8.3.2 A Modular Shell System

Figures 8.1 and 8.12 show examples of shells made by a system of modular structural elements. Peters et al. [1] called the system Polylithic. The system consists of only one single shell element shown to the left at Fig. 8.13.

The basic element consists of curved pearl-chain spines along the three edges and a pearl-chain arch at the middle. A textile mould is fixed to the spines on which the light concrete membrane can be cast. If you require a large number of equal elements for example for a structure as the one shown at Fig. 8.12, it could pay to produce a more permanent mould.

You can create a permanent mould by first casting a prototype element as explained above and then use that as a form for casting a final mould. You can make that for

Fig. 8.12 Modular shell structure by Peters et al. [1]

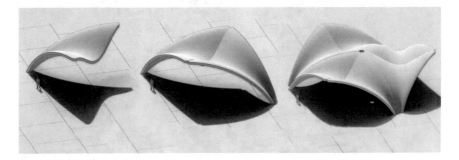

Fig. 8.13 Modular shell system. Peters et al. [1]

example as a light concrete block that you treat with a strong surface layer for example of a fibre-reinforced mortar densified with waterglass.

You can assemble two basic elements to a folded arch as shown at the middle of Fig. 8.13. Six basic elements can be assembled to a three point supported dome shell as shown to the right of Fig. 8.13. By combining domes, arches, and single elements, you may create shell structures covering a large area as shown at Fig. 8.12.

You could create simple connections between the basic elements for example by bolting steel plates that also serve as anchorage plates for the prestressing wires of the pearl-chain spines.

This allows you to create a slit between the elements that can serve to light the interior.

You may for example create the slit by inserting a piece of concrete between the shells to keep them apart. You can then connect the anchor plates in two neighbour shells with a bolt through a hole in the piece.

If you stress the bolt, you can post tension the concrete piece so that it can transfer tension as unloaded compression and serve as a prolongation of the pearl-chains in the edges of the two neighbour shells.

This means that the pearl-chains of more shell elements can act as a continuous spine in the assembled shell structure.

At the same time, the concrete piece can protect the bolt against fire.

The shells can then appear lighter and at the same time, they can become easier to erect. Figure 8.14 gives an impression of how this may appear.

This is only one example out of an infinite number of possible modular shell structure solutions.

Another way to let in natural light could simply be to substitute parts of the light concrete with windows.

Fig. 8.14 Modular structure with slits between the elements. Peters et al. [1]

8.4 Hyperbolic Shells

8.4.1 Hyperbolic Shells

Hyperbolic paraboloids also called parabolic saddle surfaces have been quite popular for in-situ cast concrete shells. The reason is that this simple double curved surface may provide a natural distribution of forces and that it contains two sets of straight lines. This means, that you can create inexpensive moulds for casting shells of this shape from straight wooden boards. Architects like for example Antonio Gaudi [6] and Félix Candela [7] investigated the static possibilities of these shell structures and showed how they can be applied for a great variety of architectural expressions.

8.4.2 Hyperbolic Element Shells

It is tempting to think that you can design hyperbolic shells from a number of equal elements all with the same twist measured as increase of inclination per unit length.

Imeri [5] investigated the possibilities and soon realized that twisted elements as those to the left in Fig. 8.15 would be of different shape although they all have the same twist and they all have an equal quadratic projection on a horizontal plane.

She realized that she could create a saddle shell shown to the right of Fig. 8.15 from trapezoidal elements. However, most elements would be different, because the arches across the saddle shape are parabolas. If you replace the parabolic arches with circles, you may design an approximate hyperbolic saddle shell from only a limited number of different trapezoidal elements.

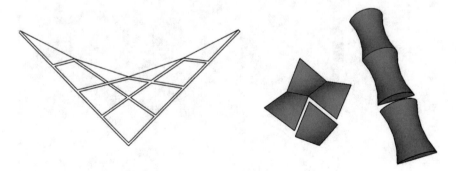

Fig. 8.15 Hyperbolic shells [5]

8.5 Tubes

8.5.1 Tubular Structures

Tubular structures may have many purposes such as pipes for transport of fluid or gas, sewers, tubes enclosing and protecting conveyors, tubes enclosing and carrying footways, roads, subways, pneumatic post systems or even pneumatic train transport systems.

An example is the tube bridge structure shown in Fig. 8.16. It is seen from the inside in Fig. 8.2. Here, the tube not only serves to cover and protect an internal footway, but also serves as a beam. The perforated sides of this post-tensioned pearl-chain tube transfer shear between top and bottom of the beam, which makes it possible to place it on pillars, so that this new element of infrastructure can be elevated above an existing developed area. The shear in the tube also transfers torsion, so that the tube can have a curvature in the horizontal plane between the supports.

Designers making tubes applied for transport have to fulfill a number of common functional requirements. The tube should protect the interior against outer effects from the climate or water pressure. Especially, the tube should be able to withstand the effect of a tube fire. Tube fires are usually violent, because heat can only escape in the longitudinal direction. This gives rise to jet fires with high temperatures and high-speed propagation.

Fig. 8.16 Super-light tube bridge by Herdal et al. [2]

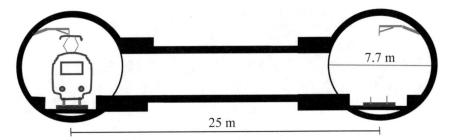

Fig. 8.17 Drilled tunnel under Great Belt in Denmark

8.5.2 Tunnels

It is tempting to design a tubular tunnel of high-strength concrete, because it has a strength to resist an outer mechanical load, and it is dense and therefore able to resist an outer water pressure. However, high-strength concrete is susceptible to explosive spalling and that in combination with powerful tunnel fires can give rise to critical damages.

8.5.3 Drilled Tunnels

If the tunnel is circular, heating also gives rise to thermal compressive stress at the inner surface, which further increases the risk of explosive spalling.

We have experienced this effect for example at fires in the Great Belt tunnel in Denmark (Fig. 8.17) and the Channel tunnel between Britain and France. See more about tunnel fires and explosive spalling in Hertz [8].

Tunnel walls made of high-strength concrete therefore require a protection against fire by for example an insulating lining for example of mineral wool.

A super-light tunnel profile solves this problem.

By placing the high-strength concrete at the outer surface, and combine it with a light concrete, the high-strength concrete can provide the denseness, and the inner light concrete can protect and stabilize the strong.

A tunnel profile with this type of cross-section could be made of elements mounted behind the head of the drilling machine in a continuous process.

8.5.4 Immersed Tunnels

A traditional immersed tunnel like the upper one in Fig. 8.18 has meter thick walls and a heavy reinforcement to sustain the bending caused by the outside water pressure.

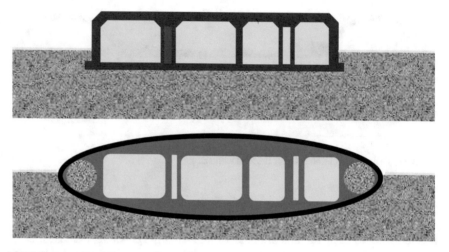

Fig. 8.18 Immersed tunnels: Traditional and super-light

The thick walls also provide the tunnel with a weight required to keep it in place at the seabed.

A super-light immersed tunnel can for example have a cross-section as the bottom one in Fig. 8.18. Here, the principle of the cross-section is the same as suggested for the drilled super-light tunnel with a dense high-strength concrete at the perimeter and a light concrete at the interior. However, the elliptical shape means that the vertical hydrostatic pressure is larger than the horizontal. You can then provide the cross-section with vertical interior walls for example made of ordinary concrete to sustain the part of the vertical water pressure that is larger than the horizontal and design the elliptic high-strength concrete to sustain a compression equal to the horizontal water pressure.

This tunnel profile saves concrete, reinforcement, and CO_2 in production. It is therefore light and easier to tow. In order to immerse the profile to the seabed, you can provide it with cavities, which you can fill with sand. Because the forces are all in compression, you can save reinforcement.

In total, the super-light immersed tunnel requires considerably less concrete and reinforcement, and it will therefore be more sustainable.

8.5.5 Floating Tunnels

Immersed tunnels are preferable, where the seabed is even and not too deep.

However, often you have conditions, where this is not the case.

For example, the Norwegian west highway should pass fjords with a depth of 1300 m, and here a floating tunnel may be desirable.

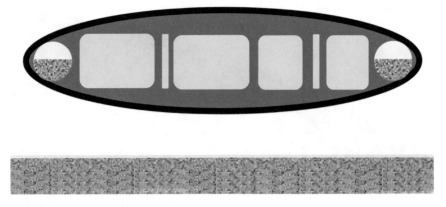

Fig. 8.19 Floating super-light tunnel

You may also have a rocky seabed, where it is possible to immerse the tunnel at certain parts, but where it will be very costly to provide an even foundation at other parts.

Here a combined solution may be an alternative, where you apply the same profile in the full length, but where it is floating in parts, where you cannot support it by an even seabed.

The proposed immersed elliptic tunnel design from 8.5.4 is therefore prepared for application as a floating tunnel (Fig. 8.19). In 8.5.4, we explained how some cavities could be filled with sand. Here, we can adjust the degree of filling to balance a desired weight of the tunnel, so that it can float and be stable for all load conditions.

Usually, it will pay to design a floating tunnel with a curvature in the horizontal plane and provide it with a longitudinal post tensioning as a long pearl-chain.

This ensures that it can counteract a sideward pressure from streaming water.

When viewed from above, the entire tunnel then acts as an arch in compression, or as a tension chain, depending on the direction of the water flow.

8.6 Floating Structures

8.6.1 Floating Structures

For floating bridges, it is also relevant to apply a curvature in the horizontal plane as mentioned for floating tunnels in Sect. 8.5.5. Bridge elements can be post-tensioned as a pearl-chain, resting on a series of caissons, and the entire bridge can hereby serve as a horizontal compression arch or tension tie for lateral pressure of streaming water.

An example is shown in Fig. 8.20.

Floating concrete structures are not something new.

Fig. 8.20   Floating bridge project. Drawing by Lund [9]

According to Pliny the Elder [14], Emperor Claudius ordered a high concrete ship to be built at Pozzuoli, where there is plenty of fly ash as explained in Sect. 1.1.2. A huge ship made for transport of Caligula's obelisk at St. Peter's Square was reused as a permanent mould. The floating concrete building was sailed and immersed as a part of the tower structure for the lighthouse at the harbour at Ostia. Archeologists have now identified the place near the Fiumicino airport of Rome.

In modern times, the Norwegians have built huge floating concrete structures as for example the Troll A platform. With a height of 472 m and weight of 684,000 tonnes, it was the tallest structure to be moved, when they towed it 200 km to the Troll oil and gas field north-west of Bergen in 1995.

You can find many other examples of caissons of concrete. You can also find numerous examples of houseboats made of concrete by specialized companies or as self-made structures.

Usually, they make a floating concrete hull and place a house of low weight on top.

A friend of one of the authors made such a houseboat, and called for advice, because the hull was leaking. When the author came to the place after a week, the concrete hull had re-sealed itself due to delayed hydration of the cement.

Experience show that such concrete structures in general are easy to maintain and repair when compared to similar floating structures of steel and wood, which are susceptible to rust and rot and need to be repainted.

Figure 8.21 shows a simple example of a floating building project by Jensen et al. [10]. They used a number of super-light flat elements with high-strength concrete at the outside and light concrete at the inside like the structure of the super-light floating tunnel in Sect. 8.5.5. First they placed a number of these elements in the shape of a capital G and joined them with a water-tight mortar. Then they post-tensioned them together as a pearl-chain.

Then they aligned several G-shaped profiles and post-tensioned them together across.

Fig. 8.21 Floating restaurant building. Drawing by Jensen et al. [10]

This made a long two-storey building, where the basement could float as a hull. Columns kept the three deck levels apart.

8.6.2 Sub-Water Cast Floating Structures

Figure 8.22 shows the principle of casting a caisson for the floating bridge shown in Fig. 8.20 by Lund [9].

Fig. 8.22 Caisson for floating bridge. Drawing by Lund [9]

She placed an inflated ring-shaped balloon at the surface of the sea carrying a submerged water filled permanent textile mould with the shape of a caisson.

She provided the mould with carbon fibres as reinforcement and with channels to be cast out with light and strong concrete.

By casting the textile mould under water, she managed to counteract the mould pressure to be an absolute minimum required to maintain the shape of the caisson. This means that she subjected the permanent textile mould to a minimum of stress. She could therefore design it using a minimum of materials and releasing a minimum of CO_2.

After casting a caisson, the water inside is pumped out. It can then float, and the ring-shaped balloon can be re-used for casting the next caissons.

The principle of combined application of textile mould for sub-water casting of light concrete seems to be an extremely beneficial and sustainable solution that can be applied for many structures at sea such as pier foundations, walls for fish farms, and hulls for floating buildings and ships.

8.6.3 Concrete Ships

During the First World War, President Woodrow Wilson ordered a flotilla of 24 concrete merchant ships to compensate for the lack of steel [11]. Twelve of them were built by different companies such as San Francisco Shipbuilding Company and by the River Jacksonville Concrete Shipyard.

In 1942, the U.S. Maritime Commission for the same reason contracted McCloskey and Company of Philadelphia to build twenty-four concrete ships [12].

Many other concrete ships have been made. For example, Pier Luigi Nervi constructed a 400-tonne coaster and a private ketch of concrete [13].

It is obvious, that textile moulds and the principle of super-light concrete structures with a dense high-strength concrete at the outer surface and a light concrete at the interior as shown in Fig. 8.23, is beneficial for building of concrete ships.

With this technology, you can create the main structures light, which means that you save material, money and CO_2. In addition, you can produce curved and double curved shapes without application of expensive moulds.

Where previous larger concrete ships have suffered from being heavy at all parts of the hull, the super-light technology means, that you can place inexpensive and CO_2 free ballast exactly where you want it to stabilize a light hull.

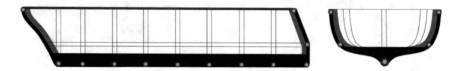

Fig. 8.23 A super-light ship

These benefits mean that you can also make a super-light ship more stable, maneuverable, and seaworthy than previous concrete ships.

References

1. Peters LS, Bekakos G, Mitrovgenis E (2020) Polylithic—a super-light concrete structure. Super-light project 2020–07. Department of Civil Engineering, Technical University of Denmark, 40p
2. Herdal RQ, Ali MM, Husarciková M, Kilic M, Calderón JD (2020) Space-time bridge. Super-light project 2020–10. Department of Civil Engineering, Technical University of Denmark, 29p
3. Ottersland J (2018) Super-light spatial structures. MSc project. Department of Civil Engineering, Technical University of Denmark. 120p.
4. Calatayud AD (2020) Design of a concrete construction to cover the Elsinore freeway. MSc project. Department of Civil Engineering, Technical University of Denmark, 168p
5. Imeri V (2019) A super-light shell structure. MSc project. Department of Civil Engineering, Technical University of Denmark, 146p
6. Lorenzi MG, Francaviglia M (2010) Art and mathematics in Antoi Gaudi's architecture: "La Sagrada Família." J Appl Math 3(1):125–146
7. Alanís, EDA (2008) Félix Candela 1910–1997 The mastering of boundaries. Taschen, 96p
8. Hertz KD (2019) Design of fire-resistant concrete structures. ICE Publishing, Thomas Telford Ltd., 254p. London. ISBN: 9780727764447
9. Lund KOA (2018) Floating concrete foundation with textile fiber bag. MSc project. Department of Civil Engineering, Technical University of Denmark, 122p
10. Jensen D, Josefsen MT, Jensen MR, Homann MH, Marquardsen T, Chatti Y (2015) KDY—Tuborg Havn. Super-light project 2015–02. Department of Civil Engineering, Technical University of Denmark, 64p
11. Stilwell B (2020) The US Navy built 12 concrete ships for World War I. https://www.wearet hemighty.com/articles/the-us-navy-built-12-concrete-ships-for-world-war-i
12. De Graaf J (2020) Concrete ships of World War I and II. https://www.thefactsite.com/concrete-ships-facts/
13. Desideri P, Nervi PLjr, Positano G (1979) Pier Luigi Nervi—A cura di Paolo Desideri, Pier Luigi Nervi jr, Giuseppe Positano. (In Italian) Zanichelli Editore, Bologna, 215p
14. Pliny the Elder (79) Naturalis Historia (natural history) Book 36, Clause 70. Dansih translation by Jabob Isager. Forlæns Publishing Company 2018, 412p

Chapter 9
Structural Detailing

Abstract Some structural details of Super-light structures are explained. The theory of reinforcement anchorage is provided, and design and full-scale tests of concrete hinges for pearl-chain arches are shown. Furthermore, a potential for easy disassembly and reuse of structural elements like the SL-decks by using a weak mortar in joints is discussed.

9.1 Anchorage

9.1.1 Anchorage Basics

In principle, anchorage failure of a reinforcing bar in concrete can occur in two ways:

(1) Bond failure, where the bar is pulled out of a round hole, and
(2) Splitting failure, where a longitudinal crack develops along the bar to the surface of the concrete structure or perhaps to other bars.

Bond strength is a property of the concrete and the corrugation of the bar.

Splitting strength depends on the geometry and strength of the cross-section.

The anchorage capacity is a minimum of bond- and splitting strength.

Abrams [1] made a pioneering work testing anchorage of a large number of reinforcing bars with varying cover thicknesses in various concrete specimens.

Goto [2] showed that when you pull a corrugated reinforcing bar embedded in concrete, shear will occur in the concrete next to the bar in planes through the bar axis. Figure 9.1 shows to the left this shear τ at the bar surface.

Since shear is compression and tension, and since tension strength is smaller than compression strength, cracks will radiate from the bar when the tension stresses exceed the tension strength of the concrete. Goto showed by tests, that these conical cracks radiate at 45° from the bar. Figure 9.1 also shows such cracks at the bar to the left.

When the cracks are formed, you cannot transfer shear any more in the concrete next to the bar. You may still be able to increase the anchored force if the structure can react against it with inclined compression stresses as also shown to the left in Fig. 9.1.

K. D. Hertz and P. Halding., *Sustainable Light Concrete Structures*, Springer Tracts in Civil Engineering, https://doi.org/10.1007/978-3-030-80500-5_9

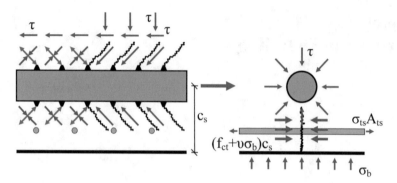

Fig. 9.1 Anchorage of reinforcement

This requires that the concrete can resist the inclined compression at the bar surface, where this compression is largest. If not, you get a bond failure. It also requires that the structure can resist the part τ of the inclined compression that radiates from the bar. If not, you get a splitting failure.

9.1.2 Splitting Strength

Tepfers [3] showed in his thesis and later in papers like [4], that a required radiating compression on a reinforcing bar may give rise to a longitudinal splitting crack along the bar to the surface of the structure. The right side of Fig. 9.1 shows a cross-section with such a longitudinal crack. The first author developed in part 2 of his thesis [5] and in a later paper [6] a method for calculation of splitting strength.

If the shear stress at the surface of a bar of radius R is τ, the surrounding structure should resist a radial compression stress of the same magnitude τ, after the conical cracks are developed. This radiating compression stress will give a hydrostatic pressure of τ in the reinforcing bar.

Figure 9.2 shows that this hydrostatic pressure is equal to a compression force 2τR over a diameter of the anchored bar per unit length of the bar. Considering equilibrium at a vertical cross-section through the centre line of the bar, you see that the surrounding structure should resist a tension of τR at each side of the bar. This means that the structure in the cover of thickness c_s to the nearest surface should resist a tension force of τR per unit length of the bar.

The right side of Fig. 9.1 shows a cross-section of a bar and forces in the cover resisting formation of the splitting crack. A main contribution to this resistance is the tensile strength of the concrete f_{ct} over the cover thickness c_s.

We often apply a cross-reinforcement with area A_{ts} per unit length in order to resist this force and codes of practise often prescribe the use of it. However, if we utilize the full tensile strength of that, we cannot take the tensile strength of the concrete into account, because the concrete cracks long before the reinforcement reaches its

Fig. 9.2 Hydrostatic
stresses in a bar

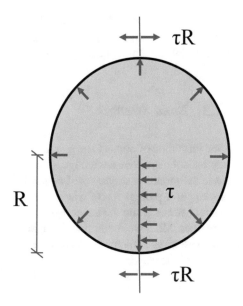

strength. In most cases, the tensile strength of the concrete is larger than the tensile strength of the cross-reinforcement. This means that, if we want to consider the cross-reinforcement in addition to the concrete tensile strength, it can only contribute with a tensile stress of $\sigma_{ts} = 31.5$ MPa, which it will have at a strain of $\varepsilon_s = 150 \cdot 10^{-6}$, where the concrete cracks. ($\sigma_{ts} = \varepsilon_s E_s$).

Very often, we would like to calculate the splitting capacity at a bearing for example at the end of a beam or slab element. In such a zone, we may have a vertical bearing stress σ_b from the reaction force. The reaction usually gives a transverse compression equal to the Poisson ratio ν of the concrete times the bearing stress σ_b. We can add this compression $\nu\sigma_b$ to the tensile strength f_{ct} of the concrete, because the splitting force should first counteract the compression before it can use the tensile strength.

Above, we have seen that the tensile strength of the structure in the cover thickness should resist τR, where τ is the shear at the surface of the anchored bar. We therefore find the anchorage force per unit length of the bar as

$$2\pi \, R\tau = 2\pi \left(A_{ts}\sigma_{ts} + c_s(f_{ct} + \nu\sigma_b)\right)$$

The effect $2\pi c_s \, \nu\sigma_b$ of the bearing stress σ_b explains why it is possible to apply anchorage lengths of only 55–70 mm at the end of many prefabricated concrete slabs.

These small anchorage lengths are required if slab elements from two sides should rest on the same wall with a thickness of 150–200 mm. Element factories have previously shown this by full-scale tests, because the empiric rules in different textbooks and codes of practise would typically require 10 times longer anchorage lengths.

The method for anchorage calculation presented above therefore represents a major improvement for assessing the anchorage capacity and for optimal design of the material consuming bearing zones of super-light elements.

9.1.3   Bond Strength

After formation of conical shear cracks as shown to the left in Fig. 9.1, the structure may be able to resist anchorage shear transfer by inclined compression stresses. These compression stresses will be largest near the surface of the bar. Bond failure happens at a pullout force, where these inclined compression stresses exceed the concrete strength and the anchorage force pulls the bar out of a round hole.

In Sect. 3.2.2, we found that the distribution of forces and stresses at an ultimate limit state gives a maximum resistance to load and deformation.

If the angle between the bar axis and the direction of the ultimate inclined concrete stress σ_{cau} is called α, the transferred shear stress τ at the bar surface will be

$$\tau = \sigma_{cau} \, \sin \alpha \, \cos \alpha = \frac{\sigma_{cau} \, \sin 2\alpha}{2}$$

From that, we see that a structure obtains the largest shear capacity and pullout strength for a given strength of the concrete, if the compressive stresses are inclined 45° degrees to the bar axis as also presumed in Sect. 9.1.2.

If you recall that the inclined compression in the concrete is conical around the bar, you may see that the ultimate compressive stresses in the concrete must the ultimate strength of the concrete with hindered lateral expansion.

Because for each radiating lamella, where you consider the inclined compression, there will be neighbour lamellas with the same compression, so that sideward expansion is not possible.

Since the biaxial concrete strength with hindered lateral expansion is about 1.3 times the monoaxial compression strength f_{cc} as shown to the right of Fig. 9.3 according to [7, 8], you may presume that $\sigma_{cau} = 1.3 \, f_{cc}$ and

$$\tau = \frac{1.3 \, f_{cc} \, \sin(2*45)}{2} = 0.65 \, f_{cc}$$

provided that the corrugations on the bar surface can transfer the full ultimate shear to the concrete.

In order to investigate the shear transfer from bar to concrete, the Nordtest foundation supported the first author in making a test method called the cuff test shown to the left in Fig. 9.3. A reinforcing bar is cast in the centre line of a concrete specimen shaped as a half standard cylinder prolonged in a 45° cone.

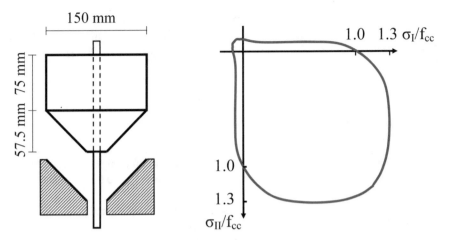

Fig. 9.3 Cuff test for bond strength and biaxial strength of concrete

The specimen is placed in a steel bearing block—the cuff—with a conical cavity provided with a layer of artificial rubber (neoprene) to distribute an effect of small irregularities.

At first, the test series comprised 280 specimens with different shapes of the reinforcing bars and diameters varying from 8 to 25 mm exposed to temperatures from 20 to 400 °C.

The test series soon proved, that an ordinary deformed reinforcing bar as the one shown in Fig. 9.1 has a bond strength of $\tau = 0.65\,f_{cc}$ as estimated theoretically above to be the maximum possible [5, 6]. The test series also proved the validity of the expression, when the concrete quality varies for example according to an increased temperature.

For plain round bars, you cannot obtain the same bond. The deformations at the surface of these bars are so small that they can lose their bond by an elastic deformation of the concrete. This has often been demonstrated at lectures by knocking the bar of the specimen on a desk.

The concrete block loses its bond by the chock and moves typically 150 mm down, where it again grips at the bar. You can repeat this up and down about 50 times before the inner hole in the concrete becomes so worn that the elastic bond is permanently lost.

The elastic bond strength of round bars were shown to be at least $\tau = 0.15\,f_{cc}$.

For some bars, rust may increase this value, but it is not recommended to rely on more.

Later, a new study was made, where the bond was investigated of twisted wires applied for prestressing lines [9]. Here we found a safe value of the bond to be $\tau = 0.25\,f_{cc}$.

The test series showed that the same expression was applicable for varying diameters of the wires and for varying concrete qualities.

The good agreement between test results and theoretical obtained values meant that the cuff test was not needed for anything else than documenting the formulas. The cuff test has therefore not been introduced as a possible new ISO standard test method.

9.1.4   Anchorage Summary

You can estimate the anchorage capacity as a force F_a per unit length of a reinforcing bar of diameter D with a cover thickness c_s from the centre line to the nearest surface in a concrete structure.

F_a is the smallest of the splitting capacity F_s and the bond capacity F_b.
Consider a concrete with compressive strength f_{cc}.
The tensile strength f_{ct} is typically about $0.1 \cdot f_{cc}$.
The Poisson ratio v is typically between 0.1 and 0.2.
In a reaction zone, you may have a stress σ_b across the cover thickness.
You may have a cross-reinforcement of area A_{ts} per unit length.
From that, you get a splitting capacity per unit length of the bar of

$$F_s = 2\pi \, (A_{ts} 31.5 \, \text{MPa} + c_s(f_{ct} + v\sigma_b))$$

and a bond capacity of

$$F_b = \pi \, D \, k \, f_{cc}$$

The factor k depends on the shape of the bar:

- For deformed bars, you can obtain the maximum theoretical value $k = 0.65$.
- Tor prestressing lines and other wires $k = 0.25$.
- For round bars $k = 0.15$.

You can find more information about anchorage at high temperatures for fire safety design in [10].

9.2   Separable Joints

9.2.1   Lime Mortar Joints

For thousands of years, masons have applied hard-burnt bricks joined with lime mortar for walls, aches, and vaults. The lime mortar transferred compression and served to glue the bricks into a composite material. It ensured dense, heavy, and sound insulating structures. The relatively weak lime mortar also allowed builders to pull down old structures for example by knocking the walls sideways. After that, the hard-burnt bricks were mostly intact, and it was possible to clean them by knocking residual lime mortar off.

Masons could then re-use the bricks in new structures. Builders could apply the debris of lime mortar as a filling material in roads or buildings, or as aggregate in concrete structures. The debris could also be grinded into gravel and applied in new mortar.

When masons introduced cement-based mortar, brickwork became stronger, but also more difficult to pull down and to clean and re-use the bricks.

From this experience, we can learn deliberately to apply the weakest possible lime mortar for joints in order to open an easy possibility for reusing the joined elements.

We usually construct super-light elements with a strong and heavy concrete of typically 55 MPa at the bearings of the elements to make them robust. Here we also introduce forces and anchor possible reinforcement, and in addition, the weight of the heavy concrete does not contribute much to the moment distribution. The builder can then design the whole structure so that the mortar of the joints should as far as possible only transfer distributed compression stresses. Lime mortar is suitable for this purpose. The weak lime mortar with small stiffness will in addition contribute to distribution of bearing stresses.

A good example of a design, which is suitable for lime mortar joints, is the beamless super-light decks in Sect. 4.3. Here, we have designed the blade connection in Fig. 4.22 to transfer reaction by distributed compression stresses in the mortar. We have also deliberately placed the connections where we do not need transfer of moments as illustrated in Fig. 4.21. The beamless super-light decks are also an example of a system of elements of standard dimensions. We can apply this system in many buildings and it has therefore a potential for being dismantled and re-used [11].

As discussed in more detail in Sect. 11.1.1 and shown in Fig. 11.6, you cannot decide that your new component should be re-used at end of life. However, you can decide to re-use a previously designed component and harvest the benefits of a low resource consume and CO_2 emission or make your new component more attractive by making it re-usable. If your structure was part of a building system with standardized dimensions and joints such as the beamless super-light decks, it becomes more likely that it can be reused and more likely that you can find a suitable used component for your structure.

9.3 Hinges and Weak Zones

9.3.1 Concrete Hinges

Hinges are applied in many structures to control the boundary conditions of the static system. Usually, the assumption is that the construction can rotate freely and without any resistance at the location of the hinge, and this may be a good conservative approach in many cases. However, in reality, there exist many types of hinge designs, where you must expect a rotational resistance.

For a pearl-chain arch or vault, a concrete hinge is an obvious choice, see Fig. 9.4. Concrete hinges were widely used in arch bridges from the beginning of the twentieth century. In the last part of the nineteenth century, Jules Dupuit was a pioneer in the use of different types of hinges to control the thrust line of masonry arches [12]. However, it was not until 1907 that the French engineer Augustin Mesnager thought of designing a hinge made of concrete [13]. The Mesnager hinge was made for concrete arch structures. The concept was simply to reduce the cross section height of the arch structure at the zone of the hinge. That zone is referred to as "the throat". Mesnager applied crossing steel bars in the throat to ensure the integrity. The famous builder Eugene Freyssinet read the work of Mesnager and developed the concrete hinge further for use in arches. He implemented it into many of his large constructions [14].

Freyssinet found that the crossing bars in the throat were not necessary when used in arches, despite of the reduced cross sectional height. The reason was that arch structures have a significant normal force through the hinge, and where the normal stresses approach the throat zone, they will bend in a bottleneck shape. This direction change causes a confinement and two-dimensional compression of the concrete in the throat, which increases the capacity. Nevertheless, Freyssinet did apply the crossing steel bars in his building to be on the safe side. Furthermore, the direction change of the normal stresses causes a splitting force in the zone next to the throat, and this would later be the focus of several studies by other researchers.

As Freyssinet began to apply concrete hinges, other engineers and builders would follow, and many great arch bridges from that time utilize the concrete hinge. Some well-known examples by Robert Maillard can still be found today in Switzerland [15, 16].

Fig. 9.4 Section cut of typical concrete hinge with crossing steel bars in the throat

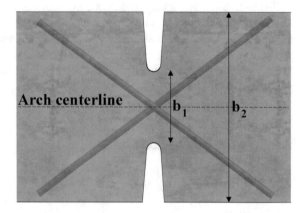

Nowadays, concrete hinges are mostly applied in bridge piers [17], but the reintro-duction of arches by using the pearl-chain principle, may also make a reintroduction of concrete hinges relevant.

9.3.2 Tests of Concrete Hinges

Several full-scale tests have been performed at DTU on concrete hinge specimens to verify their application in pearl-chain arches. All tests were to fracture. Tests specimens with different hinge throat sizes were tested with a large level of normal force corresponding to a low 30 m fully loaded pearl-chain bridge. Such low arch would have a very high level of normal force, which was the reason for testing. The test setup was a specially developed four-point bending with a constantly applied normal force. The results of the response of hinges with crossing steel bars is shown in Fig. 9.5 for three ratios between throat height b_1, and full cross section height b_2. The hinge with the biggest throat of $b_1/b_2 = \frac{1}{2}$ was not promising for application in pearl-chain arches due to cracking outside the throat. More details can be found in a paper by the authors [18]

The present guidelines for the geometry of the hinges and their reinforcements are originating from research by Leonhard and Reimann [19]. They showed that it was possible to create a universal concrete hinge response curve, by which the behaviour of any properly designed hinge could be anticipated.

Their analytical findings are verified by their own tests and tests by others [20–22].

The basic assumption of the universal hinge response theory is that there is cracking in the hinge throat, and that it is possible to relate the crack length to the level of normal force and moment in the hinge. The horizontal axis of the universal hinge curve is the rotation α divided by a parameter K expressing the strain distribu-tion, and the vertical axis is the parameter m expressing moment related to normal force and throat height

Fig. 9.5 Results of hinge responses to loading for different geometries

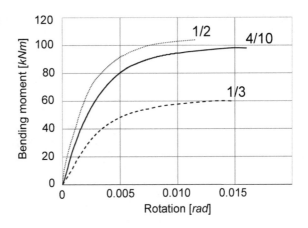

$$\frac{\alpha}{K} = \frac{1}{(1-2m)^2}$$

$$K = \frac{8 \cdot N}{9 \cdot b_1 \cdot w \cdot E_0}$$

$$m = \frac{M}{N \cdot b_1}$$

E_0 is the stiffness of the concrete in the throat and w is the width of the throat area.

Leonhardt and Reimann determined the above equations, and hereby, they could set some limits for when the hinge throat would start to crack (m = 1/6), and when the crack had reached half the throat (m = 1/3). Furthermore, they argued that a hinge should never be designed to carry load beyond the m = 1/3 limit.

The test results from our tests of concrete hinges were compared to the universal curve from Leonhardt and Reimann, and they showed that hinges with a large normal force would still fit to the universal hinge curve, see Fig. 9.6. Later, hinges with carbon fibre bars crossing in the throat were tested in full-scale as well. Results of those tests are also shown in the figure. The benefit of using carbon fibre bars instead of steel is that they cannot corrode, and hence requirements to concrete cover layers can be ignored. The CFRP-bar hinge tests gave result with a similar close fit to the universal curve [23].

It is recommended to apply hinges with the ratio $b_1/b_2 = 1/3$ in order to be sure that the crack will develop in the throat. You may then calculate the moment distribution in the structure and the rotation α of the hinge, by initially considering the hinge ideal with no moment as a conservative assumption.

Then you may calculate the parameter m relating the moment M to the normal force N and the throat height. You can then check that m is within the recommended limits of 1/3 for the ultimate limit state ULS and 1/6 for the serviceability limit state SLS.

Fig. 9.6 Test results fit to universal hinge response curve

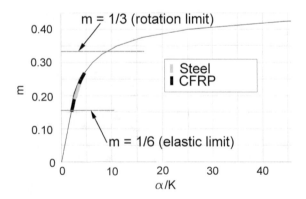

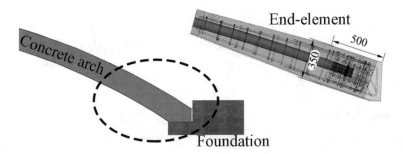

Fig. 9.7 Position and section of end-element in pearl-chain

When using concrete hinges in the supports of pearl-chain arches, you must consider whether the post-tensioning cable should be anchored before or after the hinge. If the cable runs through the hinge throat, then the post-tensioning force will add to the total normal force in the hinge and reduce m.

An alternative to the Mesnager hinge is to apply a specially designed support-element, as the end-element in the pearl-chain. Such element was used in full-scale tests to fracture of pearl-chain arches, see Fig. 9.7. The purpose of such end-element is to create a space to anchor the post-tensioning cable, and spread the local force to the whole cross-section. The transition from arch to foundation is similar to an unreinforced concrete saddle bearing. This is possible, since it will always be in compression.

The structure can rotate in the unreinforced connection between arch and foundation, if a mortar is poured in between the two when erecting the arch [24]. The stiffness and thickness of the mortar has an influence on the rotational resistance of the joint. During the full-scale tests, it was observed that cracks would initiate in the mortar layer between arch and foundation, and rotation would occur [25].

9.3.3 Weak Zones Guiding Forces and Locating Fracture

When concrete hinges are applied, you can assume that you have created a weak spot in the construction. If you use a Mesnager hinge with $b_1/b_2 = 1/3$, which is the common geometry, the bending stiffness will be 27 times smaller than that of the adjacent structure. Depending on the structure, you can therefore anticipate that cracking will initiate in the concrete hinge throat first.

This means that you, e.g. in relation to arch structures hinged at the supports, can quickly assess the status of the arch by looking for cracks at the hinges. Furthermore, the arch thrust line is guided through a hinge, and this may help you to control where the forces are.

Since concrete hinges are not ideal, the structural response will be in between the behaviour of ideal hinges, and that of fixed connections. Figure 9.8 shows an

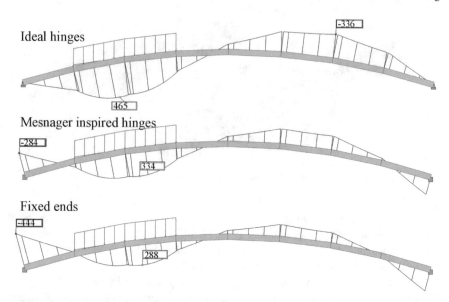

Fig. 9.8 Moment distribution of loaded arch with and without concrete hinges

example of an arch with two hinges at the supports compared to an arch with fixed ends.

You may choose to create an arch without hinges, but hereby you may not know the location of where the structure will start to crack. The structure must then form four plastic hinges before failure, unless the loading will cause a direct compressive failure in the arch cross-section. Furthermore, the very rigid fixed arch will be susceptible to settlements of the foundations, temperature variations, creep etc.

If you choose a structure with two hinges, one at each support, you have provided the arch with the ability to withstand vertical settlements of the foundation. However, horizontal settlements of the foundation may still cause some increase in the bending moment. Still the arch is also susceptible to creep and temperature variations.

A three-hinged arch, with an additional hinge at the top, is statically determinate, and therefore easy to approach as an engineer. Furthermore, it is not susceptible to creep, temperature variations or settlements. Unfortunately, the three-hinged static system will not provide any "reserve strength" if the foundation moves outwards due to the arch load, and therefore this type of system should only be applied where the soil conditions are good or you apply tension ties as explained in Sects. 6.2.5, 6.2.6, and 7.2.3.

The authors full-scale tested two adjacent pearl-chain arches. Each was supported by two saddle bearing hinges as illustrated in Fig. 9.7. The test showed a number a warning signs before failure. The arches were loaded with a critical load around the ¼ point of the span. First, the cracking at the hinges initiated and propagated. Then significant deformations were visible for the arch structure, and finally, two plastic hinges formed and created a mechanism at failure load.

9.3.4 Super-Light Concrete Hinges

As explained in Chap. 3, the super-light principle allows the engineer to guide the forces, where they are most useful by application of concrete of different density and stiffness.

You can benefit from this by designing a concrete hinge as shown in Fig. 9.4, but filling out the cavities on both sides of the reduced cross-section with a light concrete as for example a foam concrete of density 300 kg/m^3 (see Sect. 2.2.2).

In practise, you can cast these inserts first, and apply them as permanent parts of the mould.

Doing so, you can make the mould for the hinge cheaper, and you obtain an insulation of the hinge, protecting the concrete as well as any reinforcement in the hinge against fire.

In addition, you can hinder stiff objects such as bottles etc. to fall into the groove, and you obtain an even surface of the hinge, which again can serve for example as a mould for casting a filling material in a sandwich arch as explained in Sect. 7.3.

References

1. Abrams DA (1913) Test of bond between concrete and steel. Urbana, University of Illinois. Eng Exp Stat Bull 71:240
2. Goto Y (1971) Cracks formed in concrete near a reinforcing bar due to bond and transverse cracking. J Am Concr Inst Proc 68(4):244–251
3. Tepfers R (1973) A theory of bond applied to overlapped tensile reinforcement splices for deformed bars. Thesis. Publication 73:2, Division of Concrete Structures, Chalmers University of Technology, Gothenburg, 328p
4. Tepfers R (1979) Cracking of concrete cover along anchored deformed reinforcing bars. Mag Concr Res 31(106):3–12
5. Hertz KD (1980) Armeringsståls forankring ved høje temperaturer. (Anchorage of reinforcing steel at high temperatures) in Danish. Report 138 and part 2 of Ph.D. thesis on fire exposed concrete structures, Institute of Building Design, Technical University of Denmark, 103 p
6. Hertz KD (1982) The anchorage capacity of reinforcing bars at normal and high temperatures. Mag Concr Res 34(121):213–220
7. Kupfer H, Hilsdorf HK, Rusch H (1969) Behavior of concrete under biaxial stresses. ACI J 66:656–666
8. Neville AM (1977) Properties of concrete. The Pitman Press, Bath, p 687p
9. Hertz KD (2005) Vedhæftningsstyrken af spændliner ved brandpåvirkning (Bond strength of prestressing lines exposed to fire) (In Danish). Report SR 05–12 Department of Civil Engineering, Technical University of Denmark, 10p
10. Hertz KD (2019) Design of fire-resistant concrete structures. ICE Publishing, Thomas Telford Ltd, p 254p
11. Halding PS, Hertz KD (2020) Design for disassembly of super-light structures. RILEM spring convention 2020
12. Holzer SM, Veihelmann K (2015) Hinges in historic concrete and masonry arches. Proc ICE Eng Hist Heritage 168(2):54–63. https://doi.org/10.1680/ehah.14.00019
13. Mesnager A (1907) Experiences sur une semi-articulation pour voutes en Béton armé. Annales de Ponts de Chaussees 2:180–201

14. Fernandez-Ordonez D (2018) Eugene freyssinet: I was born a builder. 28. Dresdner Brücken-baussymposium, Technische Universität Dresden
15. Billington D (1984) Building bridges: perspectives on recent engineering. Ann N Y Acad Sci 424:309–324
16. Billington D (2000) The revolutionary bridges. Sci Am 283(1)
17. Morgenthal G, Olney P (2015) Concrete hinges and integral bridge piers. J Bridge Eng 21(1). https://doi.org/10.1061/(ASCE)BE.1943-5592.0000783
18. Halding PS, Hertz KD, Schmidt JW (2014) Concrete hinges. In: Proceedings of the Iass-slte 2014 symposium
19. Leonhardt F, Reimann H (1965) Betongelenke—Versuchsbericht, Vorschlaege zur Bemessung und konstruktiven Ausbildung. Deutscher Ausschuss Fuer Stahlbeton, pp 1–34
20. Base GD (1965) Tests on four prototype reinforced concrete hinges. Cement and concrete association—research report, pp 1–28
21. Marx S, Schacht G (2010) Concrete hinges—historical development and contemporary use. In: Proceedings of the 3rd international fib congress and exhibition, USA
22. Tourasse M (1961) Essais sur articulation Freyssinet (Experiments on Freyssinet hinges). Ann Institute Technique du Batiment et des Travaux Publics 40(57):62–87 (in French)
23. Halding PS, Schmidt JW, Musachs S (2021). Static behaviour of concrete hinges with CFRP bars. Future article in pipeline
24. Halding PS, Hertz KD, Viebæk NE, Kennedy A (2015) Assembly and lifting of pearl-chain arches. Proc Fib Symp 2015:185–186
25. Halding PS, Hertz KD, Schmidt JW, Kennedy BJ (2016) Full-scale load tests of pearl-chain arches. Eng Struct 131:101–114

Chapter 10
Sustainability

Abstract The combination of light aggregate concrete and ordinary concrete decreases the environmental impact from super-light structures. The chapter provides an overview of CO_2-emissions from different materials and building parts to compare with super-light structures. The comparisons are made realistic by complying with actual requirements to sound insulation.

10.1 Environmental Impact

The Environmental impact of our human activities has become more and more in focus. The emission of CO_2 and other greenhouse gases causes global warming and climate changes, which have become visible as extreme weather. Temperatures increase, water levels increase, the number of hurricanes, forest fires, and floodings increases.

A climate debate is taking place, where people require solutions from politicians and technicians. It is especially required that the present generation should not transfer their CO_2 problems to the next, but solve them instead (Fig. 10.1).

An early example of a study on these matters was a Ph.D. project by Andersen [1, 2], which was an answer to a question, only few had asked at the time. He found a level for energy consume of 500–1000 kWh per m^2 for dwellings equal to 150–300 kg CO_2 per m^2, which corresponds well to modern calculations.

This work combined with knowledge about ancient building technology inspired the first author to develop new methods for constructing light sustainable concrete structures.

He invented Super-light structures in 2008, and Pearl-chain structures in 2009.

In 2010, he and the Technical University of Denmark formed a start-up company called Abeo Ltd. with a group of three students from Copenhagen Business School.

The name means, "abandon the old stuff" in Latin to mark a new beginning in structural design.

The new technologies made it possible and economically feasible to produce structures with less material consume and the same load-bearing capacities, improved sound-insulating properties, and a better fire-resistance than traditional structures.

© The Author(s), under exclusive license to Springer Nature Switzerland AG 2022
K. D. Hertz and P. Halding., *Sustainable Light Concrete Structures*, Springer Tracts in Civil Engineering, https://doi.org/10.1007/978-3-030-80500-5_10

Fig. 10.1 Masai Mara Kenya. *Photo* KD Hertz

In cooperation with architects, consulting engineers, and producers, the first author, his students and the Abeo Company developed the first suggestions for super-light alternatives to actual building projects. They calculated the CO_2 emission for structural parts and building processes for the lifetime of a number of projects. These calculations demonstrated CO_2 savings between 20 and 50% compared to ordinary concrete structures, 80% compared to steel structures and up to 50% compared to timber structures, when the calculations compared the ratio of CO_2 release and lifetime of the buildings.

In addition, the smaller consume of materials means less pollution from production of materials and components and from erection of buildings. Application of clean non-toxic materials such as lime, clay, sand, and water and new possibilities using less carbon for heating them in the production processes means a further reduction of pollution.

The new structures are therefore more climate-friendly and sustainable than traditional building technologies, and this is why the start-up company won the price as "Best Danish Early-stage Cleantech Company" in 2010.

The same year it won the "Clean Tech Open Global Ideas Competition" in San Francisco (Fig. 10.2). Byrne [3] declared in The New York Times:

> *Abeo thinks BIG for its Super-light Structures. For a company which has only been in existence since June this year, Danish start-up Abeo has been racking up the awards for rethinking an everyday construction material—concrete.*

The word BIG was a reference to the Bjarke Ingels Group architects, which was one of the cooperating companies that had just opened an office in New York City.

In 2013, the EU-commission gave Abeo Ltd. a prize as "Europe's Best Innovative Spin-out Company" (Fig. 10.3).

Fig. 10.2 Trophy for the world championship in clean technology won by the first author's start-up company in 2010. *Photo* KD Hertz

Fig. 10.3 Alexander Wulff from Abeo Ltd receives the prize as Europe's Best Innovative Start-up Company in 2013. *Photo* EU-Commission

However, in 2010 neither building industry nor politicians were willing to invest in clean technology and CO_2 savings. Abeo had to close down in 2016, and the producer took over the company that today sells super-light deck elements in a number of countries.

10.1.1   Systematic CO_2 Assessment

You can often find many different values for the CO_2 release of a product or a material. Some producers find it tempting to give the assessment of their own product a beneficial treatment, and sometimes they forget important details and contributions to the calculation. Furthermore, you cannot apply fixed values for all products and materials. Very often, you find considerable variations in manufacturing of a product or a material from producer to producer depending on the processes applied and on transport of raw materials to the location of production. You can also find considerable variations for the same product or material made by the same producer at different times. The producer may for example implement changes to make the product more sustainable and of course, you should consider this in the assessment. Else, the development has been in vain.

This means that you cannot make a fixed table with reliable definitive values of CO_2 for building materials and components. Instead, in each case, you have to consider the processes involved for the particular producer at the time of production, and even then, data are scattered due to random variations in the processes and due to variation in the nature of the raw materials. As shown in Fig. 10.4, each process gives a product and for that, you summarize the CO_2 footprints. They consist of one part for the raw material plus one part for energy such as heat, electricity or mechanics plus one for human operation plus one of removing and treating waste products.

If the process gives a by-product, you may consider, if someone can apply it. In most cases, the by-product is considered to be CO_2 free by the receiver, and it should

Fig. 10.4 Process

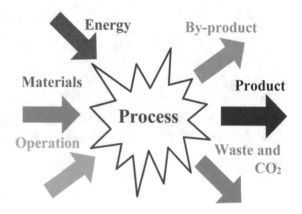

not be subtracted from the account of the process. For example, a contribution to district heating of a city is valuable with respect to CO_2, when it replaces heat from a coal driven plant in wintertime. However, you cannot take this CO_2 saving into account, if the extra heat cannot be applied by the city in summertime, or if the plant already has a fossil free energy production.

In this book, we are concerned about the impact of a building on the climate.

The CO_2 emission damaging the climate comprises that of production of materials, the building process, and the impact from the end of life stage. The latter includes demolition of the building and processing materials as waste.

The end of life stage is especially important, because without that you may think that you have a low CO_2 impact, where you in reality only obtain to store your old CO_2 for a while.

Later, it is released damaging the climate for the next generations in addition to their own CO_2 contributions.

References like the standard EN 15804 and Herrmann [4, 5] therefore explain this in detail. They emphasize the importance of the end of life stage and group the processes involved in main classes A, B, and C. Each class is subdivided as for example A is divided into A1-A5 (Fig. 10.5).

It is obvious that you should relate the CO_2 emission of the life cycle including the end of life phase to the lifetime of a structure or material. It is also obvious that a comparison of climate impact of different building structures, components, or materials can only make sense if they fulfill the same functional requirements for the actual application.

For example, you can only compare emission of a carbon fibre reinforcing bar with that of a steel bar, if they both have the same strength, and they are placed in the structures so that the steel bar will not be damaged by rust and the carbon fibre bar will not be damaged by fire exposure.

Some calculations include a benefit when combustible materials are burned at the end of life stage and replace fossil fuels like coal in power plants producing district heating. Likewise, oven heat from production of cement or light aggregates may

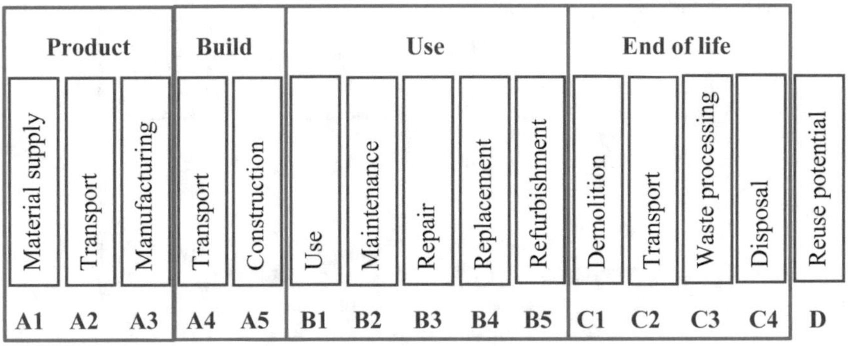

Product			Build		Use					End of life				
Material supply	Transport	Manufacturing	Transport	Construction	Use	Maintenance	Repair	Replacement	Refurbishment	Demolition	Transport	Waste processing	Disposal	Reuse potential
A1	A2	A3	A4	A5	B1	B2	B3	B4	B5	C1	C2	C3	C4	D

Fig. 10.5 Life Cycle phases according to CEN

contribute to district heating. This is only relevant, if the power plants are based on fossil fuels, and if you can foresee that, they still will be that, at the time for end of life perhaps 50 years from now. In most industrialized countries, this is not the case, and such contributions are not included in the numbers presented in this book.

A potential for reuse will prolong the lifetime of a material or component and reduce the CO_2 impact related to the lifetime. The standard [4] suggests adding this as a phase D. You should not consider this on beforehand, when you apply a material or component for the first time in a structure and calculate the related CO_2 impact.

At this early stage, you do not know whether an industry will reuse them at the end of life of your building structure 50 years or more after construction.

Furthermore, a future building industry will be less willing to reuse your old material or component, if you have already taken the benefit on your CO_2 account.

If you for example apply a component that in a life cycle of phase A-C gives an impact of 500 kg CO_2 for a lifetime of 50 years, you have to report an impact of 10 kg CO_2 per year.

If it were accepted that you instead presumed that someone would reuse your component in a future building for another 50 years, you would only report the half impact of 5 kg CO_2 per year. In such a system, the future builder should also report 5 kg CO_2 per year as you have done for a new component.

This means that the future builder has no incentive to reuse your old component and no payment for the extra work and logistic of finding it and fitting it into the new building.

We therefore recommend implementing a possible reuse phase D at the beginning of the life cycle (Fig. 10.6). Here, you can be sure about how much you intend to

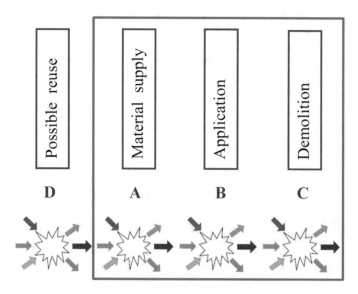

Fig. 10.6 Lifetime impact and absorption

reuse in your own construction, and you will have a clear benefit from doing so.

Recycled steel is a good example of an implemented technology that gives rise to a considerably reduced CO_2 impact that industry already takes into account as a phase D at the beginning of the life cycle and not at the end of it.

The Ökobaudat database refers to the GaBi database [6] for actual CO_2 numbers. Other databases and sources of information like [7] from the University of Bath and [8] from VTT, give values for the CO_2 impact, which are only slightly different, when you transfer their properties to numbers with the same meaning. This represents no serious problem since the preconditions such as the nature of the sub processes, the efficiency, and the time and place for deriving the numbers may be different.

The problem becomes serious, if you apply the values without including a possibility for adjusting them according to well-documented improvements or variations of the processes in question. If you do not do that, companies have no incitement for changing their way of doing things, and your calculation method will be a hindrance for development.

This means that you can apply general data from the databases assessing the CO_2 impact. However, as soon as you have more precise and well-documented data for an actual product, you should apply them instead.

In the following, we present estimates of the CO_2 emission from (phase A) production, (phase B) use, and (phase C) demolition of materials and products based on accessible knowledge.

The values presented include equivalents for other greenhouse gases such as methane and perfluorocarbons (PFC's) etc.

For comparison, we also present ideal values for how little CO_2 emission we can reach, if all sub processes become as CO_2 neutral as possible.

This indicates that we can expect considerable changes of the impact in the future.

It also shows how important it is that the basis for assessment of CO_2 impact can follow the changes, so that a too conservative assessment method does not hinder the positive development (Fig. 10.7).

10.2 CO_2 from Materials

10.2.1 Concrete

The building industry accounts for a large part of the global CO_2 emission, and cement production alone causes about 6% of the emission of the Earth. This amount is increasing while building industry is developing mainly in China, but also in India, and Africa [9, 10]. We therefore have to address technologies like those of this book, by means of which, we can reduce cement consume and create structures with application of less concrete.

However, when we make comparisons between alternative building materials and methods in order to choose optimal solutions, it is important to be able to estimate the

Fig. 10.7   Super-light deck element at Krøyers Plads. Architect Cobe

CO_2 production for the possibilities in a correct manner. This gives us a problem as explained in Sect. 10.1. You cannot look up the CO_2 development from something like cement production as a single number in a table. This property varies considerably from factory to factory, and it changes over time depending on the attention the industry pays to developing environmentally friendly production methods.

For a traditional concrete, you produce cement in an oven with a large heat loss fired by fossil fuel, and you do nothing to replace some of the cement with for instance natural pozzolana. Furthermore, you apply ordinary local stone as a heavy aggregate. For this traditional heavy concrete, you may expect to emit:

0.14 kg CO_2 per kg concrete with strength 55 MPa.

The cement counts for 90% of this emission, since 1 kg concrete contains 0.14 kg cement with 0.9 kg CO_2 per kg cement. We can estimate this roughly in accordance with [7, 11, 12].

You can then find that the cement contributes with 0.9 kg CO_2 per kg cement · 0.14 kg cement per kg concrete = 0.126 kg CO_2 per kg concrete. Other contributions (mainly transport and processing of aggregates) count for about 0.010 kg CO_2 equal to the residual 10%. In total 0.126 + 0.010 = 0.14 kg CO_2 per kg concrete.

This value is also comparable with [7, 13–15].

If the concrete has a smaller compressive strength of 25 MPa, the cement content is only about 0.10 kg per kg, but the CO_2 from transport etc. is the same. You then emit. 0.9·0.10 + 0.010 = 0.090 + 0.010 = 0.100, where [15] estimates 0.094 kg

CO$_2$ per kg concrete.

0.09 kg CO$_2$ per kg concrete with strength 25 MPa.

The chemical process in the cement production counts for approximately 50% of the CO$_2$ emission [16], this is from burning limestone to calcium oxide combined with quarts, aluminium or iron [17].

$$CaCO_3 \rightarrow CaO + CO_2$$

You may often reduce 25% of the cement with **fly ash** and get:
0.75*0.9·0.14 + 0.010 = 0.095 + 0.010 = 0.105 kg CO$_2$ pr kg 55 MPa concrete and.
0.75*0.9·0.10 + 0.010 = 0.068 + 0.010 = 0.078 kg CO$_2$ pr kg 25 MPa concrete.

0.11 kg CO$_2$ per kg 55 MPa concrete with 25% fly ash or pulverized ceramic

0.08 kg CO$_2$ per kg 25 MPa concrete with 25% fly ash or pulverized ceramic

Of this respectively 0.095 or 0.068 kg CO$_2$ per kg concrete is from cement production and of that 50% (equal to respectively 0.048 or 0.034 kg CO$_2$ per kg concrete) is from the chemical process.

If you create a new green cement replacing 35% of it with pulverized burned clay, you may reduce the CO$_2$ emission from the cement with 25–30%.

As mentioned in Sect. 1.1.2 this is the oldest recipe for concrete practised by the Phoenicians more than 3000 years ago.

The Phoenicians fired their ovens with wood. In our time, we could do the same by means of any bio fuel, so that the heating emits much less CO$_2$. Alternatively, we could heat our burned limestone ovens with electricity, as we usually already do with our ceramics ovens and glass ovens, and we could make the electricity from wind, water, or sun power plants.

Likewise, the producer can power the mechanical processes electrically.

This means that we can reduce the CO$_2$ emission to the one released chemically by production of the burned limestone CaO:

0.05 kg CO$_2$ per kg fossil free produced concrete 55 MPa

0.03 kg CO$_2$ per kg fossil free produced concrete 25 MPa

In addition, cement hydration products gradually reabsorb atmospheric CO$_2$ through the process of carbonation. [17] show quantities for this and how the CO$_2$ uptake depends on the CO$_2$ concentration, the density of the concrete, the strength class, exposure conditions, additions, and coatings. You have not only CO$_2$ uptake by Calcium Hydroxide but also by the other minerals in the hydrated concrete like:

$$Ca(OH)_2 + CO_2 \rightarrow CaCO_3 + H_2O$$

$$(3CaO \cdot 2SiO_2 \cdot 3H_2O) + 3CO_2 \rightarrow 3CaCO_3 \cdot 2SiO_2 \cdot 3H_2O$$
$$(2CaO \cdot SiO_2) + 2CO_2 + xH_2O \rightarrow 2CaCO_3 + SiO_2 \cdot xH_2O$$
$$(3CaO \cdot SiO_2) + 3CO_2 + xH_2O \rightarrow 3CaCO_3 + SiO_2 \cdot xH_2O$$
$$(3CaO \cdot Al_2O_3 \cdot 6H_2O) + 3CO_2 \rightarrow 2Al(OH)_3 + 3CaCO_3 + 3H_2O$$

Pade and Guimaraes [18] show that approximately 25% of the concrete carbonize during a 100 years lifetime, and further 25% when you demolish it.

Lo et al [19] show that application of 25% fly ash gave a marginal increase of carbonation.

In a life cycle **after carbonation**, you can then estimate the CO_2 release of ordinary concrete with cement from coal fired ovens to be:

$0.5 \cdot 0.126 + 0.010 = 0.078$ for ordinary 55 MPa concrete after carbonation and

$0.5 \cdot 0.090 + 0.010 = 0.055$ for ordinary 25 MPa concrete after carbonation and

If all mechanical processes are made electric so that 0.010 is reduced to 0 and 25% fly ash or ceramics are applied, and the ovens are powered fossil free so that the CO_2 is only made chemically, and this is reduced by 50% carbonation you get ultimate reduced values as

$$0.5 \cdot 0.048 = 0.024 \text{ for } 55 \text{ MPa and}$$
$$0.5 \cdot 0.034 = 0.017 \text{ for } 25 \text{ MPa}$$

This means that

0.024 kg CO_2 per kg 55 MPa fossil free with 25% fly ash/ceramics after demolition

0.017 kg CO_2 per kg 25 MPa fossil free with 25% fly ash/ceramics after demolition

This release is identical to the one made by the Phoenicians and Romans. It should therefore be possible for us to do the same today. It is a matter of investing in a changed production.

10.2.2   Light Aggregate Concrete

Light aggregates are usually produced by burning granulated clay in a rotary kiln.

When you heat with coal to 1150 C with a 150% heat loss from the process, you emit

$0.104 \text{ kg } CO_2 \text{ per MJ} \cdot 1.5 \cdot 1150 \text{ C} \cdot (1 \text{ kJ}/(\text{kg C})) \cdot 1 \text{ MJ}/(1000 \text{ kJ}) = 0.179 \text{ kg } CO_2$

per kg light aggregate.

Fig. 10.8 Light aggregate concrete blocks

A 600 kg per m³ light aggregate concrete (Fig. 10.8) has typically 400 kg expanded clay, 100 kg sand and 100 kg cement per m³. For a 900 kg per m³ light aggregate concrete the numbers are 600, 150 and 150 kg per m³. This means that the cement and light aggregate content is proportional to the weight. The CO_2 footprint per kg is therefore the same

From cement $0.9 \cdot 100/600 = 0.150$ kg CO_2 per kg concrete

$+$ From aggregate $0.179 \cdot 400/600 = 0.120$ kg CO_2 per kg concrete

$+$ From sand $0.003 \cdot 100/600 = 0.001$ kg CO_2 per kg concrete

In total, the emission is 0.270 kg CO_2 per kg light aggregate concrete. [15] gives 0.30 kg CO_2 per kg for 700 kg/m³ and 0.24 kg CO_2 per kg for 1600 kg/m³.

0.27 kg CO_2 per kg concrete with expanded clay.

Typically, you replace 20% of the cement with fly ash. Doing so, you get:
From cement $0.9 \cdot 0.8 \cdot 100/600 = 0.120$ kg CO_2 per kg concrete and in total

$0.120 + 0.120 = 0.240$ kg CO_2 per kg light aggregate concrete.

0.24 kg CO_2 per kg light $-$ aggregate concrete with 20% fly ash.

This is a typical material, and the number is the same as found by VTT for light-aggregate concrete blocks of density 500–1600 kg/m³ [8].

The Danish light clay aggregate factory, and no doubt also other factories in the world, is now using waste as fossil free fuel replacing 80–90% of the applied coal. The last 10% of the applied carbon is embedded in the clay and cannot be replaced. With these aggregates, you emit 0.120 from cement $+ 0.15*0.120$ from light aggregate $= 0.120 + 0.018 = 0.138$ kg CO_2 per kg

0.14 kg CO_2 per kg concrete with light – aggregate heated by waste and 20% fly ash.

As mentioned in 10.2.1 approximately 50% of the CO_2 release from cement comes from the chemical processes and the rest from energy applied. This means that if the cement is produced fossil free and you apply almost fossil free produced aggregate or pumice, you only get a contribution from cement of $0.5 \cdot 0.120 = 0.060$ kg CO_2 per kg concrete. In total

0.08 kg CO_2 per kg fossil free produced light aggregate concrete with 20% fly ash.

Even light-aggregate concrete of relatively high densities of about 1800 kg/m^3 carbonize significantly faster than similar heavy concrete according to for example [19] and [20]. Carbonizing may be a problem for oxidation of steel reinforcement [21], but for many of the applications of this book you do not use reinforcement of steel in parts of the structure, where light aggregate concrete is applied.

Light-aggregate concrete with density less than 1200 kg/m^3 is very porous and therefore has a larger internal surface exposed to CO_2 from the air and a more rapid carbonizing time than the heavy dense concrete qualities [22].

Where we previously in this book could assess the carbonizing of a heavy dense concrete to be 25% during its lifetime and further 25% at demolition, we can presume it to be at least 66% in total and very often 100% in total for the light concrete qualities of about 600 kg/m^3 as we apply in super-light structures.

This means that we can assess the emission from cement to be between 0 and $0.34 \cdot 0.060 = 0.020$ kg CO_2 per kg carbonized light aggregate concrete, and in total

0.02 – 0.04 kg CO_2 per kg carbonized fossil free light
aggregate concrete with 20% fly ash

10.2.3   Pumice Concrete

If you apply volcanic material like pumice as light aggregate for a 600 kg per m^3 concrete, (with 400 kg pumice, 100 kg sand and 100 kg cement per m^3 replaced 20% by fly ash), you emit from cement $0.9 \cdot 0.8 \cdot 100/600 = 0.120$ kg CO_2 per kg concrete + From aggregate 0 kg CO_2 per kg concrete + From mechanics and transport as for heavy concrete 0.010 kg CO_2 per kg concrete $= 0.130$ kg CO_2 per kg pumice concrete of 600 kg/m^3 and proportional the same per kg for 900 kg per m^3 concrete.

0.13 kg CO_2 per kg 600 − 900 kg/m³ concrete with pumice and 20% fly ash

If the pumice should be transported by sea from Iceland to Denmark it costs 16 g CO_2 per 1000 kg pumice per km, and you get for 1800 km an addition of $0.016 \cdot 1800 \cdot 400/600 =$

0.019 Kg CO_2 per kg pumice Concrete and in total

0.15 kg CO_2 per kg 600 − 900 kg/m³ concrete with pumice and 20% fly ash in DK

10.2.4   Foam Concrete

We now look at a foam concrete made by mixing foam into a mortar of cement and sand.

Mix design corresponds to [23]. 600 and 900 kg per m³ foam concrete with 300 and 330 kg cement per m³. Both have replaced 20% cement by fly ash. We assess the other contributions as similar to those for other types of concretes. The emission is:

$$0.90 \cdot 0.8 \cdot 330/900 + 0.010 = 0.264 + 0.010 = 0.274$$

0.28 kg CO_2 per kg foam concrete of 900 kg/m³ with 20% fly ash

$$0.90 \cdot 0.8 \cdot 300/600 + 0.010 = 0.360 + 0.010 = 0.370$$

0.37 kg CO_2 per kg foam concrete of 600 kg/m³ with 20% fly ash

A fossil free production will then lead to

0.13 kg CO_2 per kg fossil free foam concrete of 900 kg/m³ with 20% fly ash

0.18 kg CO_2 per kg fossil free foam concrete of 600 kg/m³ with 20% fly ash

Also for this, you may expect more carbonation than for a heavy concrete and you end up with between 0 and $0.34 \cdot 0.18 = 0.061$ kg CO_2 per kg fossil free foam concrete

0 − 0.06 kg CO_2 per kg carbonized fossil free foam concrete with 20% fly ash

10.2.5   Aerated Concrete

Aerated concrete is produced by aluminium powder that chemically produces bubbles with the cement. It is usually autoclaved as a part of the production process. Estimates of the CO_2 emission can be found in different sources, where different methods of production presumably cause the different values [8, 15]:

> **0.51 kg CO_2 per kg aerated reinforced concrete block 433 kg/m^3**
>
> **0.34 kg CO_2 per kg aerated reinforced concrete 480 kg/m^3**

10.2.6   Steel

Steel production consists of several processes each giving rise to release of CO_2 partly from heating and partly from the chemical reactions taking place in the material (Fig. 10.9).

At first, you make coke from coal by heating it above 600 °C in absence of oxygen.

Then, you make pig iron (also called crude iron) by melting at 1600 °C the raw materials of iron ore Fe_2O_3, coke C and lime $CaCO_3$ and impurities such as ferrous sulphide FeS in a blast furnace.

$$2Fe_2O_3 + 6C + 3O_2 \rightarrow 4Fe + 6CO + 3O_2 \rightarrow 4Fe + 6CO_2 \quad \text{and}$$
$$2FeS + 2CaCO_3 + C \rightarrow 2CaS + 2FeO + 2CO_2 + C \rightarrow 2CaS + 2Fe + 3CO_2$$

Fig. 10.9 Firth of forth bridge by Fowler and Baker 1889. *Photo* KD Hertz

The pig iron has a carbon content of about 4% and silica and other impurities making it too brittle for application in structures.

Then, the primary steel process reduces the content of carbon to about 1%.

$$C + O_2 \rightarrow CO_2$$

For that, you apply a Basic Oxygen Furnace or an Electric Arc Furnace, which has replaced the old Bessemer furnace, because they have better pollution control systems.

In this process, you also want to remove a surplus of oxygen by adding aluminium.

$$4Al + 3O_2 \rightarrow 2Al_2O_3.$$

Then in the so-called secondary steel process, you adjust the composition of the molten steel.

All these processes of traditional steel production lead to a relatively large CO$_2$ emission mainly caused by chemical reactions. You can assess the release as

3.00 kg CO$_2$ per kg virgin steel for the traditional production methods

2.80 kg CO$_2$ per kg virgin steel and wrought iron for modern production methods

This is according to for example [7, 15]. Germeshuizen and Blom [24] and others indicate a possibility of reducing the CO$_2$ emission from steel productions drastically by application of hydrogen replacing carbon as reducing agent.

$$3Fe_2O_3 + 9H_2 \rightarrow 2Fe_3O_4 + H_2O + 8H_2 \rightarrow 6FeO + 3H_2O + 6H_2 \rightarrow 6Fe + 9H_2O$$

They estimate that this may reduce the CO$_2$ emission to

0.18 kgCO$_2$ per kg virgin steel for a <u>future</u> hydrogen based steel production

The steel industry still has to develop and implement this. One of the problems for this is to avoid excessive hydrogen in the steel that makes it brittle. Therefore, you may consider this only to be a possible future value.

10.2.7 Recycled Steel

Hammond and Jones [7] also estimate a value for recycled steel as 0.47 kg CO$_2$ per kg, where [15] estimates 0.73 kg CO$_2$ per kg probably due to different production methods. These differences and differences in the fraction of reused steel that is applied on average leads to different values for the steel usually applied in Germany and UK.

With the average content of recycled steel in UK, [7] estimate

1.46 kg CO_2 per kg steel with average recycled content in UK

With the average content of recycled steel in Germany, [15] estimates

1.00 kg CO_2 per kg steel profile with average recycled content in Germany

Applying the same share of recycling as used today, you get for a future recycled hydrogen based steel $0.18 \cdot 1.46/2.80 =$

0.094 kg CO_2 per kg steel for future recycled hydrogen based steel,

If a steel profile is galvanized, this protection increases the CO_2 impact. According to [15] you emit

1.85 kg CO_2 per kg galvanized steel profile

Stainless steel is a collective name for a number of alloys with different materials. Their CO_2 impacts are therefore varying according to the content and the methods of production. You may find values like [7, 8]

3.78 kg CO_2 per kg stainless steel according to Ruuska [8]
6.15 kg CO_2 per kg stainless steel according to Hammond and Jones [7]

Reinforcing steel
Reinforcing steel with an average recycled content gives according to [15]

0.75 kg CO_2 per kg reinforcement

Typically, you apply 100 kg reinforcement per m^3 concrete increasing the density from 2300 to 2400 kg per m^3. This means that a typical value for reinforced concrete can be found by adding the CO_2 impact from the reinforcement to that of the concrete and relate the sum to the new larger density.

Using an average value for concrete of 0.13 kg CO_2 per kg, this will give you an addition of

$100 \, kg \cdot 0.75 \, kg \, CO_2$ per $kg/2400 \, kg - 0.13 \cdot (2400 - 2300)/2400 = 0.031 - 0.005$

$= 0.03$ kg CO_2 addition per kg concrete from 100 kg reinforcement per m^3

10.2.8   Timber

Some politicians believe that wood and timber structures (Fig. 10.10) represent a solution to the climate crisis, because trees absorb CO_2 when they grow. However, the trees release the CO_2 again, when they stop growing and start to rot, burn, or are eaten by insects.

When you apply wood in a structure, you have to prevent it from rotting or burning by taking special precautions and often by impregnation or painting with poisonous chemicals at regular intervals. The more CO_2 you store in wood, the more comprehensive is the task that you hand over to the next generations, preventing the wood from rotting or burning and releasing the old CO_2.

The authors consider this behaviour to be the opposite of saving the next generation from our CO_2 problems. Instead, we claim that it is a better solution to the climate problems, if you reduce the release of greenhouse gases to a minimum and build structures with a long lifetime in order to save CO_2 from material production, transport, and building processes.

At present, timber structures give rise to CO_2 footprints from the processes of forestry, sawmill work, transport, construction, maintenance, and demolition.

In the following, we try to find representative values by comparing different sources estimating the CO_2 emission.

Most data sources give values from phase A, B and C representing the full life cycle (Figs. 10.5 and 10.6). However, the much-used reference Ökobaudat [15] gives a separate value for CO_2 emission from phase A comprising the materials, making the components, and constructing of the building. For wood structures, this becomes a negative number, since trees absorb CO_2 when they grow. For example, Ökobaudat

Fig. 10.10 Copenhagen's first Central Station by Herholdt 1863. *Photo* KD Hertz

gives a value of -632 kg CO_2 per m^3 CLT (Cross Laminated Timber) that with a density of 489 kg/m^3 gives -1.291 kg CO_2/kg CLT.

However, you release the accumulated CO_2 again when the material burns or rots.

Ökobaudat [15] includes a value for this release in phase C "End of Life" as 1.802 kg CO_2 per kg wood. We need to add this value to the negative value of phase A and B in order give the impact on the climate and to compare it to values from other sources representing phase A, B, and C. This gives a total of 0.51 kg CO_2 per kg CLT, which is comparable to values from other data sources.

Sometimes, you hear an argument, that new trees are planted and they absorb the CO_2 released by the old. However, this argument makes no sense, since the new trees also releases their CO_2.

In order to investigate this further we make a separate check of the value for phase C. We compare the value applied above for phase C with the same found by other references. Using a standard value for CO_2 release from burning of wood (for example from Quasching [25] or Engineering Toolbox [26], you get 0.108 kg CO_2 per MJ.

References like Aniszewska and Arkadiusz [27] and Alakangas [28] find that at 12% moisture, the energy released by burning is about 16 MJ/kg increasing for a smaller moisture content, where the density also decreases.

This gives a value of $16 \cdot (MJ/kg) \cdot 0.108$ kg CO_2/MJ $= 1.73$ kg CO_2/kg, which indicates that the value 1.80 kg CO_2 per kg wood for phase C from [15] comprising demolition and incineration in a municipal solid waste plant appears to be reasonable. We have therefore added this value for phase C to all values for phase A1-3 from Ökobaudat. As explained, this gives 0.51 kg CO_2 per kg CLT as shown below.

0.51 kg CO_2 per kg cross laminated timber (CLT)

Lifetime of timber structures is in general smaller than for stone structures due to rot, fire, and insects. However, reuse of unharmed timber is more common, and these two opposite effects makes it reasonable to compare CO_2 emissions for equal life times.

For ordinary timber, different sources may give slightly different values as for example.

0.31 Kg CO_2 per kg timber in [7] and 0.38 kg CO_2 per kg sawn pine timber of density 549 kg/m^3 in [15].

We apply average values

0.35 kg CO_2 per kg timber
0.52 kg CO_2 per kg laminated wood, plywood, or chipboard

10.2.9   Other Materials

Usually, you apply a number of other materials in structures or as a part of them protecting them from impact of fire or weather.

For some of them as for example aluminium, you have the same issue of recycling, as we saw for steel that virgin and recycled aluminium have quite different CO_2 emissions, and you may take an average degree of recycling into account [7].

> 12.50 kg CO_2 per kg virgin aluminium profile
>
> 2.12 kg CO_2 per kg recycled aluminium profile
>
> **10.93 kg CO_2 per kg aluminium profile with average recycling**

A data sheet at the end of the book summarises the CO_2 data given in this chapter and provides data for a number of other materials as gypsum board, brick, glass, aluminium, mineral wool and foam plastic.

10.3 CO_2 from Processes and Transport

10.3.1 CO_2 from Heating

Heating with oil, diesel, or gas gives rise to

> 0.24 kg CO_2 per kWh $= 0.24/3.6 = 0.067$ kg CO_2 per MJ energy.

Heating with coal or wood gives rise to

> 0.39 kg CO_2 per kWh $= 0.37/3.6 = 0.108$ kg CO_2 per MJ energy.

10.3.2 CO_2 from Transport

As long as we base transport on fossil fuel, you need to estimate the transport needed for raw materials, building materials and building elements.

The following average values are based on the work made by [29], and reported by [30]

> 8 g CO_2 for 1000 kg per km ship deep − sea container
>
> 16 g CO_2 for 1000 kg per km ship short sea
>
> 31 g CO_2 for 1000 kg per km barge

22 g CO_2 for 1000 kg per km rail

      (Average, 35 diesel hauled, 18 electric hauled. ECTA)

62 g CO_2 for 1000 kg per km road

602 g CO_2 for 1000 kg per km air

10.4 CO_2 from Building Components

10.4.1 Building Components with 55 dB Sound Insulation

A main hindrance for reducing CO_2 emission producing domestic building components is the consideration of sound insulation. A good sound insulation reducing noise at low frequencies requires that the separating building component contain a certains mass.

In the following, we analyze the total CO_2 emission for decks and walls fulfilling the Danish requirement of 55 dB sound insulation.

To obtain that with a massive structure of one material, you have to apply 440 kg per m^2 wall or deck structure.

In super-light deck elements (SL-decks) as the one in Fig. 10.11, you combine layers of materials of quite different stiffness and eigenfrequency such as heavy strong concrete of density 2300 kg per m^3 and a light concrete of density 700 kg per m^3. The two materials oscillate differently, and some of the sound reduces to heat.

Fig. 10.11 SL-deck element for DTU Building 130. *Photo* KD Hertz

This is the reason why you may reduce the total mass required for these components to about 340 kg per m^2.

10.4.2   Building Decks with 55 dB Sound Insulation

190 mm massive concrete deck
A 190 mm slab of 55 MPa concrete with for example five pretensioned 12.5 mm wires per m emits 437 kg concrete of 0.14 kg CO_2 per kg and 2.7 kg reinforcement per m^2. In total:

63.8 kg CO_2 per m^2 190 mm massive concrete deck

220 mm hollow-core slab with additional 55 mmtop concrete
A 220 mm hollow core deck of 55 MPa concrete has in an actual design 6 holes of diameter 147 mm per 1.2 m width. This gives a concrete mass of 311 kg per m^2 and to that 3.65 kg per m^2 from the prestressing reinforcement. Therefore, it requires a 55 mm top concrete of 126 kg per m^2, which can be made from a 25 MPa quality. This gives $44.0 + 11.3 + 2.7 =$

58.1 kg CO_2 per m^2 220 mm hollow – core slab with 55 mm top concrete

180 mm super-light SL-deck with 20 mm top concrete.
It uses the five pretensioned wires per m, 233 kg per m^2 strong concrete of 55 MPa and 58 kg per m^2 light aggregate concrete of 700 kg/m^3.
It can obtain the 55 dB sound insulation with a total weight of 340 kg per m^2.
20% of the cement is replaced with fly ash and you get 0.24 kg CO_2 per kg light concrete.
In Denmark and some other places, light aggregates are made in an 85% fossil free (FF) process. For these materials, you emit 0.14 kg CO_2 per kg light concrete.
CO_2 emission from light concrete per m^2 then becomes 58 kg · 0.24 = 13.9 kg CO_2 (FF 8.1 kg CO_2). To that, you should add 233 kg strong concrete of 55 MPa and 0.14 kg CO_2/kg (giving 32.6 kg CO_2 per m^2) + 45 kg top concrete of 25 MPa and 0.09 kg CO_2 /kg (giving 4.1 kg CO_2 per m^2) + 2.7 kg CO_2 per m^2 for the reinforcement.
In total, you emit $32.6 + 13.9 + 4.1 + 2.7 =$

53.3 kg CO_2 per m^2 180 mm SL deck with 20 mm top concrete

With 85% fossil free light aggregates, you emit $32.6 + 8.1 + 4.1 + 2.7 =$

47.5 kg CO_2 per m^2 180 mm SL deck with 85% fossil free
light aggregates and with 20 mm top concrete

175 mm CLT five-layer slab with 110 mm *top concrete*

A 175 mm five-layer CLT slab of density 489.4 kg/m^3 has a mass of 85.6 kg per m^2.

You should provide it with a 110 mm top concrete of 254 kg/m^2 in order to reach the 340 kg/m^2 that you at least should apply to reach the required 55 dB sound insulation for a composite structure.

It emits $85.5 \cdot 0.51 = 43.7$ from the CLT $+ 254 \cdot 0.09 = 22.9$ from the concrete $=$

66.6 kg CO$_2$ per m^2 five − layer CLT deck with 110 mm top concrete

10.4.3   Sand as Sound Insulator for Decks

340 kg per m^2 required for composite sound insulation has been obtained by adding sand for example to a CLT deck. However, it gave practical problems with sand moving and even penetrating the structure, when the deck was subjected to oscillations.

Therefore, the contractors known to have applied it, now use a concrete topping instead.

Nevertheless, if you calculate this solution, you emit for a five layer CLT slab of 86 kg per m^2

43.7 kg CO$_2$ per m^2 five − layer CLT with additional 254 kg

sand per m^2 (not a pragmatic solution)

If you compare this with a 90 mm pretensioned massive concrete slab of 55 MPa concrete of 210 kg per m^2 with additional 130 kg sand per m^2, the impact becomes

31.2 kg CO$_2$ per m^2 .90 mm massive concrete slab

+ additional 130 kg sand per m^2 (not a pragmatic solution)

10.4.4   Building Walls with 55 dB Sound Insulation

190 mm massive concrete wall

437 kg 25 MPa concrete of 0.09 kg CO$_2$ per kg emits 39.3 kg CO$_2$ per m^2 wall.

Two layers of 6 mm reinforcement per 250 mm in each direction give 3.55 kg reinforcement of 0.75 kg CO$_2$ per kg and this emits 2.7 kg CO$_2$ per m^2 wall.

The wall has a total weight of 440 kg/m^2, and the emission is

$$\textbf{42.0 kg CO}_2 \textbf{ per m}^2 \textbf{ massive 190 mm concrete wall}$$

Massive CLT wall
For a massive timber wall the emission would be similar to a massive timber slab.
For a massive wall or slab to obtain 55 dB sound insulation, you should apply 440 kg/m^2, and if that is CLT (Cross Laminated Timber) with a density of 489.4 kg/m^3, this requires a thickness of 899 mm emitting

$$\textbf{224.4 kg CO}_2 \textbf{ per m}^2 \textbf{ massive CLT wall}$$

You can also, for walls with 55 dB sound insulation, benefit from making composite structures such as the examples shown below.
180 mm sandwich concrete wall
With two layers of 65 mm normal concrete of 25 MPa of 299 kg/m^2 and 50 mm light concrete of 35 kg per m^2 and two reinforcing nets. The total weight is 340 kg per m^2 and that gives 55 dB since 360 kg has proven by test to give 58 dB with the same relative proportions between light and heavy concrete.
Using light aggregate concrete of 0.24 kg CO$_2$ per kg you emit 26.9 from the normal concrete + 8.4 from the light + 2.7 from the reinforcement =

$$\textbf{38.0 kg CO}_2 \textbf{ per m}^2 \textbf{ sandwich wall of 50 mm}$$
$$\textbf{light and 2} \cdot \textbf{65 mm normal concrete}$$

With 85% fossil free light aggregates with 0.14 kg CO$_2$ per kg light concrete, the result is

$$26.9 + 4.9 + 4.1 + 2.7 =$$

$$\textbf{34.5 kg CO}_2 \textbf{ per m}^2 \textbf{ sandwich concrete wall with 85\%}$$
$$\textbf{fossil free light aggregates}$$

Composite CLT walls were tested and calculated by [31].
The best structure fulfilling a sound insulation requirement of 55 dB was:
CLT wall of two slabs with a void between
The slabs are each 120 mm, and the void of 95 mm can be filled with mineral wool.
It has a minimal total weight of only 171 kg/m^2 and emits

$$\textbf{87.4 kg CO}_2 \textbf{ per m}^2 \textbf{ CLT sandwich wall of 2} \times \textbf{120 mm}$$
$$\textbf{+ 95 mm mineral wool}$$

10.4.5   Building Components with Other Requirements

100 mm massive prefabricated concrete deck (not for domestic buildings)
   230 kg concrete of 55 MPa (0.14 kg CO_2 per kg) emits 32.2 kg CO_2 per m^2 deck. In addition 3.65 kg reinforcement emits (0.75 kg CO_2 per kg) 5.3 kg CO_2 per m^2 deck.

   (3.65 kg is equal to five 12.5 mm prestressing lines per m, each consisting of 93 mm^2 of steel with density 7850 kg/m^3).

   The total weight of such deck is 234 kg/m^2. Note that this deck is not sufficient for domestic buildings in general because a sound insulation requirement of 55 dB would require a mass of 440 kg/m^2. However, it is applied for industrial buildings.

34.9 kg CO_2 per m^2 100 mm massive prefabricated

concrete deck for industry

150 mm massive concrete wall
   345 kg concrete of 25 MPa (0.09 kg CO_2 per kg) emits 31.1 kg CO_2 per m^2 wall- Two layers of 6 mm reinforcement per 250 mm in each direction gives 3.55 kg reinforcement (0.75 kg CO_2 per kg) and this emits 2.7 kg CO_2 per m^2 wall. The total weight of the wall is 349 kg/m^2.

33.7 kg CO_2 per m^2 massive 150 mm concrete wall for industry

200 mm massive concrete wall (with 58 dB sound insulation)
   460 kg of 25 MPa concrete (0.09 kg CO_2 per kg) emits 41.4 kg CO_2 per m^2 wall. Two layers of 6 mm reinforcement per 250 mm in each direction gives 3.55 kg reinforcement per m^2 wall (0.75 kg CO_2 per kg), and this emits 2.7 kg CO_2 per m^2 wall.

   The wall has a total weight of 464 kg/m^2, and gives an emission of

44.1 kg CO_2 per m^2 massive 200 mm concrete

wall with 58 dB sound insulation

220 mm SL-deck (with 58 dB sound insulation)
   The 71 kg light aggregate concrete (0.24 kg CO_2 per kg) emits 17.0 kg CO_2 per m^2 deck. The 285 kg normal concrete of 55 MPa (0.14 kg CO_2 per kg) emits 39.9 kg CO_2 per m^2 deck. The 3.65 kg reinforcement (0.75 kg CO_2 per kg) emits 2.7 kg CO_2 per m^2 deck. (3.65 kg is equal to five 12.5 mm prestressing lines per m each consisting of 93 mm^2 with steel density 7850 kg/m^3). The slab proved to give about 58 dB sound insulation, which is more than the required 55 dB. The total weight of the deck is 360 kg/m^2, and it emits $39.9 + 17.0 + 2.7 =$

59.7 kg CO_2/m^2 220 mm SL $-$ deck with 58 dB sound insulation

With 85% fossil free light aggregates the emission is $39.9 + 9.9 + 2.7 =$

52.6 kg CO_2 per m^2 220 mm SL – deck with 85%
fossil free light aggregates and 58 dB sound insulation

180 mm SL-deck
It has 58 kg light concrete that emits 13.9 kg CO_2 per m^2 deck and 233 kg normal 55 MPa concrete that contributes with 32.6 kg CO_2 per m^2 deck. Furthermore, an emission of 2.7 kg CO_2 per m^2 deck comes from 3.7 kg reinforcement.
The total weight of the deck is 295 kg/m^2, and it emits $32.6 + 13.9 + 2.7 =$

49.3 kg CO_2/m^2 180 mm SL – deck for industry

With 85% fossil free light aggregates, it becomes $32.6 + 8.1 + 2.7 =$

43.5 kg CO_2 per m^2 180 mm SL – deck with 85%
fossil free light aggregates

200 mm concrete sandwich wall (with 58 dB sound insulation)
The sandwich wall consists of two layers of 72 mm normal concrete of 2300 kg/m^3, and one layer of 56 mm light concrete of 700 kg/m^3. The light concrete contributes with 39 kg and 9.3 kg CO_2 per m^2 (0.24 kg CO_2 per kg). In addition, the normal concrete of 55 MPa weights 332 kg and emits (0.09 kg CO_2 per kg) 29.9 kg CO_2 per m^2 wall. 3.6 kg reinforcement emits 2.7 kg CO_2 per m^2 deck. (3.6 kg is equal to two nets of 6 mm bars per 250 mm with density 7850 kg/m^3). This gives a total weight of 375 kg/m^2 and approximately 60 dB sound insulation. The emission is $29.9 + 9.3 + 2.7 =$

41.9 kg CO_2 per m^2 200 mm sandwich wall of 50 mm
light and 2 · 65 mm normal concrete and 58 dB sound insulation

With 85% fossil free light aggregates, the emission is.
(0.14 kg CO_2 per kg) $29.9 + 5.4 + 2.7 =$

38.0 kg CO_2 per m^2 200 mm sandwich wall of 50 mm light 85%
fossil free concrete and 2 · 65 mm normal
concrete and 58 dB sound insulation

150 mm sandwich concrete wall (with 45 dB sound insulation)
The sandwich wall consists of two layers of 54 mm normal concrete of 2300 kg/m^3 and one layer of 42 mm light concrete of 700 kg/m^3. The light concrete contributes with 29 kg weight and 7.0 kg CO_2 per m^2 (0.24 kg CO_2 per kg). In addition, the normal concrete of 55 MPa weights 249 kg and emits (0.09 kg CO_2 per kg) 22.4 kg

CO_2 per m^2 wall. 3.6 kg reinforcement emits 2.7 kg CO_2 per m^2 deck. (3.6 kg is equal two nets of 6 mm bars per 250 mm with density 7850 kg/m^3). This gives a total weight of 282 kg/m^2 and approximately 45 dB sound insulation. The emission is $22.4 + 7.0 + 2.7 =$

32.1 kg CO_2 per m^2 150 mm sandwich wall of 42 mm

light concrete and 2 * 54 mm normal concrete and 45 dB

sound insulation for industry

With 85% fossil free light aggregates with 0.14 kg CO_2 per kg $22.4 + 4.1 + 2.7$
$=$

29.2 kg CO_2 per m^2 150 mm sandwich wall of 42 mm

light 85% fossil free concrete and 2 * 54 mm

normal concrete and 45 dB sound insulation for industry

10.4.6   Sustainability

The numbers in this chapter demonstrate how you can save CO_2 emission by application of the new technology presented in the book, when you compare components with the same functional requirements. They also show possibilities for obtaining further reductions in the future by application of improved production methods already known today.

The applied materials such as lime, clay, sand, and water are clean and non-toxic, and methods for eliminating the application of carbon for production are already implemented in some factories.

The authors therefore hope, that the technology presented in the book can contribute to a more climate-friendly and sustainable building industry, which has also been evaluated by independent price committees.

References

1. Andersen S (1979) The accumulated energy consume for building materials. Report 134, (in Danish) DTU Byg (Inst.of Building Design) 1979, 198p
2. Andersen S (1980) The accumulated energy consume for dwellings. Report 137, (in Danish) DTU Byg (Inst.of Building Design) 1980, 124p
3. Byrne C (2010) VentureBeat San Francisco The New York Times November 17, 2010
4. CEN (2012) EN 15804 Sustainability of construction works. Environmental product declarations, +A2(2019):66p

5. Herrmann H (2017) Ökobaudat Basis for the building life cycle assessment. Forshung für die Praxis Volume 11. Federal Insitute for Research on Building, Urban Affairs and Spatial Development. Bonn, 40p
6. Kupfer T et al (2019) GaBi database and modelling principles
7. Hammond G, Jones C (2011) Inventory of carbon and energy (ICE). University of Bath Version 2.0. http://www.circularecology.com/embodied-energy-and-carbon-footprint-database.html#. XbbGPuhKhaR
8. Ruuska A (2013) Carbon footprint for building products. ECO2 data for materials and products with the focus on wooden building products. VTT Technology 115. VTT
9. Benhelal E et al (2013) Global strategies and potentials to curb CO_2 emissions in cement industry. J Clean Prod 51:142–161
10. Gregg JS et al. (2008) China: Emissions pattern of the world leader in CO_2 emissions from fossil fuel consumption and cement production. Geophys Res Lett Adv Earth Space Sci 35(8)
11. NRMCA (2008) Concrete CO_2 Fact Sheet. Publication Number 2PCO2. National Ready Mixed Concrete Association
12. Pentalla V (1997) Concrete and sustainable development. ACI Mater J 94(5):409–416
13. Hertz KD (2010) CO_2 emissions from super-light structures. Department of Civil Engineering, Technical University of Denmark, 6p, 2010
14. Hertz KD, Bagger A (2011) CO_2 emissions from super-light structures. Proceedings of the IABSE-IASS Symposium "Taller, Longer, Lighter" 7p, London 20–23 Sep 2011
15. Ökobaudat (2020) Database. German Federal Ministry of the Interior, Building and Community (BMI). https://www.oekobaudat.de/datenbank/browser-oekobaudat.html
16. Lehne J, Preston F (2018) Making concrete change—Innovation in low-carbon cement and concrete. Chatham House, Royal Institute of International Affairs, 122p. https://www.cha thamhouse.org/publication/making-concrete-change-innovation-low-carbon-cement-and-con crete
17. Xi F et al (2016) Substantial global carbon uptake by cement carbonation. Nature Geoscience, vol 9. Macmillan Publishers Ltd
18. Pade C, Guimaraes M (2007) The CO_2 uptake of concrete in a 100 year perspective. Cement Concr Res 37:1348–1356
19. Lo TY et al (2007) Comparison of carbonation of lightweight concrete with normal weight concrete at similar strength levels. Constr Build Mater 22:1648–1655
20. Yingli G et al (2013) Effects of different mineral admixtures on carbonation resistance of lightweight aggregate concrete. Constr Build Mater 43:506–510
21. Lagerblad B (2005) Carbon dioxide uptake during concrete life cycle—State of the art. Report 2:2005 Swedish Cement and Concrete Research Insitute, 47p
22. Goltermann P et al (2017) Climate friendly concrete. (In Danish: Klima venlig Beton) For Region H. Department of Civil Engingeering, Technical University of Denmark. BYG R-371, 53p
23. SirContec (2010) www.sircontec.com
24. Germeshuizen LM, Blom PWE (2013) A techno-economic evaluation of the use of hydrogen in a steel production process, utilizing nuclear process heat. Int J Hydrogen Energy 38:10671–10682
25. Quasching V (2019) Specific carbon dioxide emissions of various fuels. Erneubare energien und klimaschutz, Berlin. https://www.volker-quaschning.de/
26. Engineering Toolbox (2019). https://www.engineeringtoolbox.com/
27. Aniszewska M, Arkadiusz G (2014) Comparison of heat of combustion and calorific value of the cones and wood of selected forest trees species. Lesne Prace Badawcze (Forest Research Papers) 75(3):231–236
28. Alakangas E (2005) Properties of fuels used in Finland—BIOSOUTH-project. VTT processes report PRO2/P2030/05 technical research centre of Finland, 100p
29. McKinnon A, Piecyk M (2011) Measuring and managing CO_2 emissions. Logistics Research Centre, Hariot-Watt University, Edinburgh, UK, 36p. https://cefic.org/app/uploads/2018/12/ MeasuringAndManagingCO2EmissionOfEuropeanTransport-McKinnon-24.01.2011-REP ORT_TRANSPORT_AND_LOGISTICS.pdf

30. ECTA (2011) Guidelines for measuring and managing CO_2 emission from freight transport operations. European Chemical Transport Association. Brussels, 18p. https://www.ecta.com/resources/Documents/Best%20Practices%20Guidelines/guideline_for_measuring_and_managing_co2.pdf
31. Ljunggren F (2019) Sound insulation prediction of single and double CLT panels. In: (Luleå Univ.) Proceedings of 23rd international congress on acoustics, Aachen, pp 242–248

Chapter 11
CO_2 Data

Materials

CO_2 emissions per 2020 from materials including a full life cycle with production, transport, building, application, demolition and removal.

Steel virgin from factory and wrought iron	2.80 kg CO_2/kg
Steel recycled	0.47 kg CO_2/kg
Steel profiles with average recycled content in Germany	1.00 kg CO_2/kg
Steel profile hot galvanized	1.85 kg CO_2/kg
Reinforcement with average recycled content	0.75 kg CO_2/kg
100 kg Reinforcement in 2400 kg/m^3 concrete gives an addition of	0.03 kg CO_2/kg
Cement	0.90 kg CO_2/kg
Concrete 55 MPa	0.14 kg CO_2/kg
Concrete 25 MPa	0.09 kg CO_2/kg
Light concrete (600–900 kg/m^3)	0.27 kg CO_2/kg
Light concrete with 20% fly ash	0.24 kg CO_2/kg
Light concrete with 20% fly ash and 85% fossil free aggregates	0.14 kg CO_2/kg
Aerated concrete 480 kg/m^3 on aluminium powder	0.34 kg CO_2/kg
Timber	0.35 kg CO_2/kg
CLT Cross-Laminated Timber 489 kg/m^3	0.51 kg CO_2/kg
Laminated wood, plywood, chipboard 507 kg/m^3	0.52 kg CO_2/kg
Gypsum plaster board 630 kg/m^3	0.23 kg CO_2/kg
Brick 1800 kg/m^3	0.23 kg CO_2/kg
Glass 3 mm 2500 kg/m^3	1.37 kg CO_2/kg
Aluminium profile	10.93 kg CO_2/kg
Mineral wool	1.60 kg CO_2/kg
Foam Plastic PE	6.13 kg CO_2/kg

Orign of data

Anybody can look the presented data up in references like ÖkobauDat [7] or LCA Byg [8] and you would get almost the same data from other references like Hammond and Jones [3] or Ruuska VTT [5].

© The Author(s), under exclusive license to Springer Nature Switzerland AG 2022   203
K. D. Hertz and P. Halding., *Sustainable Light Concrete Structures*, Springer Tracts in Civil Engineering, https://doi.org/10.1007/978-3-030-80500-5_11

The data represent the full life cycle including the end of life stage, which is important for the impact on the climate. End of life is especially important for organic materials like wood that absorbs CO_2 when it grows, and releases it at the end of life. Sources like [3] and [5] do therefore not present the absorbed CO_2 without the end of life contribution in their numbers. However, Using ÖkobauDat [7] or LCA Byg [8], you have to add end of life as explained in detail in [1] and [6]. For example, [7] gives a value of -632 kg CO_2/m^3 for Cross Laminated Timber CLT. With density 489 kg/m^3 it is -1.291 kg CO_2/kg. To that, you should add a value of 1.80 kg CO_2/kg for end of life, so that you get 0.51 kg CO_2/kg for CLT. This is close to the values from the other sources.

Transport

8 g CO_2 for 1000 kg per km ship deep sea container

16 g CO_2 for 1000 kg per km ship short sea

31 g CO_2 for 1000 kg per km barge

22 g CO_2 for 1000 kg per km rail

(1 km electric per 1.9 km diesel hauled. ECTA)

62 g CO_2 for 1000 kg per km road

602 g CO_2 for 1000 kg per km air

Energy

Diesel oil or Gas 0.24 kg CO_2/kWh $= 0.067$ kg CO_2/MJ.

Coal or Wood 16 $\cdot$(MJ/kg) $\cdot$ 0.108 kg CO_2/MJ $= 1.73$ kg CO_2/kg.

Building structures with **55 dB** sound insulation	
190 mm Massive concrete deck 55 MPa 440 kg/m^2	**63.8 kg CO_2/m^2**
220 mm Hollow-core + 126 kg 55 mm top concrete 440 kg/m^2	**58.1 kg CO_2/m^2**
180 mm SL-deck + 45 kg 20 mm top concrete	**53.3 kg CO_2/m^2**
180 mm SL-deck with 85% fossil free aggr + 45 kg top concrete	**47.5 kg CO_2/m^2**
220 mm SL-deck 360 kg/m^2. (makes 58 dB)	**59.7 kg CO_2/m^2**
220 mm SL-deck with 85% fossil free aggr (makes 58 dB)	**52.6 kg CO_2/m^2**
900 mm Massive CLT deck or wall 440 kg/m^3	**224.4 kg CO_2/m^2**
175 mm CLT deck 5-ply 86 kg/m^2 with **110 mm top concrete**	**66.6 kg CO_2/m^2**
Steel deck 150 kg/m^2 with **290 kg top concrete** per m^2	**303.6 kg CO_2/m^2**
190 mm massive concrete wall 440 kg/m^2	**42.0 kg CO_2/m^2**
180 mm sandwich concrete wall 340 kg/m^2	**38.0 kg CO_2/m^2**
180 mm sandwich concrete wall with 85% fossil free aggr	**34.5 kg CO_2/m^2**
CLT wall as tested 2·120 mm CLT + 95 mm air 171 kg/m^2	**87.4 kg CO_2/m^2**